ISBN 978-1-5284-0805-9
PIBN 10913186

1 MONTH OF
FREE
READING

at

www.ForgottenBooks.com

By purchasing this book you are eligible for one month membership to ForgottenBooks.com, giving you unlimited access to our entire collection of over 1,000,000 titles via our web site and mobile apps.

To claim your free month visit:

www.forgottenbooks.com/free913186

Bibliographic Notes / Notes techniques et bibliographiques

> obtain the best original
atures of this copy which
e, which may alter any of
uction, or which may
al method of filming are

L'Institut a microfilmé le meilleur exemplaire qu'il lui a
été possible de se procurer. Les détails de cet exemplaire qui sont peut-être uniques du point de vue bibli-
ographique, qui peuvent modifier une image reproduite,
ou qui peuvent exiger une modification dans la méthode normale de filmage sont indiqués ci-dessous.

☐ Coloured pages / Pages de couleur

☐ Pages damaged / Pages endommagées

☐ Pages restored and/or laminated /
Pages restaurées et/ou pelliculées

☑ Pages discoloured, stained or foxed /
Pages décolorées, tachetées ou piquées

☐ Pages detached / Pages détachées

☑ Showthrough / Transparence

☐ Quality of print varies /
Qualité inégale de l'impression

☐ Includes supplementary material /
Comprend du matériel supplémentaire

☐ Pages wholly or partially obscured by errata slips,
tissues, etc., have been refilmed to ensure the best
possible image / Les pages totalement ou
partiellement obscurcies par un feuillet d'errata, une
pelure, etc., ont été filmées à nouveau de façon à
obtenir la meilleure image possible.

☐ Opposing pages with varying colouration or
discolourations are filmed twice to ensure the best
possible image /. Les pages s'opposant ayant des
colorations variables ou des décolorations sont
filmées deux fois afin d'obtenir la meilleure image
possible.

minated /
ou pelliculée

re de couverture manque

éographiques en couleur

an blue or black) /
re que bleue ou noire)

ustrations /
is en couleur

/
nents

hadows or distortion along
e serrée peut causer de
ion le long de la marge

g restorations may appear
possible, these have been
peut que certaines pages
rs d'une restauration
e, mais, lorsque cela était
pas été filmées.

produced thanks

ınada

e best quality
n and legibility
ıg with the

ɔvers are filmed
d ending on
ıstrated impres-
ropriate. All
ıginning on the
ated impres-
with a printed

nicrofiche
ıning "CON-
ing "END"),

filmed at
ɔo large to be
are filmed
orner, left to
frames as
illustrate the

L'exemplaire filmé fut reproduit grâce à la
générosité de:

Bibliothèque nationale du Canada

Les images suivantes ont été reproduites avec le
plus grand soin, compte tenu de la condition et
de la netteté de l'exemplaire filmé, et en
conformité avec les conditions du contrat de
filmage.

Les exemplaires originaux dont la couverture en
papier est imprimée sont filmés en commençant
par le premier plat et en terminant soit par la
dernière page qui comporte une empreinte
d'impression ou d'illustration, soit par le second
plat, selon le cas. Tous les autres exemplaires
originaux sont filmés en commençant par la
première page qui comporte une empreinte
d'impression ou d'illustration et en terminant par
la dernière page qui comporte une telle
empreinte.

Un des symboles suivants apparaîtra sur la
dernière image de chaque microfiche, selon le
cas: le symbole ➡ signifie "A SUIVRE", le
symbole ▽ signifie "FIN".

Les cartes, planches, tableaux, etc., peuvent être
filmés à des taux de réduction différents.
Lorsque le document est trop grand pour être
reproduit en un seul cliché, il est filmé à partir
de l'angle supérieur gauche, de gauche à droite,
et de haut en bas, en prenant le nombre
d'images nécessaire. Les diagrammes suivants
illustrent la méthode.

2   3   1

                                                    2

MICROCOPY RESOLUTION TEST CHART

(ANSI and ISO TEST CHART No. 2)

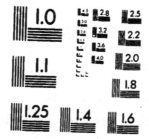

APPLIED IMAGE Inc

1653 East Main Street
Rochester, New York 14609 USA
(716) 482 - 0300 - Phone
(716) 288 - 5989 - Fax

CANADA
DEPARTMENT OF MINES
HON. MARTIN BURRELL, MINISTER; R. G. McCONNELL, DEPUTY MINISTER.

GEOLOGICAL SURVEY
WILLIAM McINNES, DIRECTING GEOLOGIST.

MEMOIR 111

No. 91, GEOLOGICAL SERIES

The Silurian Geology and Faunas of Ontario Peninsula, and Manitoulin and Adjacent Islands

BY

M. Y. Williams

OTTAWA

Fig. 1

Flower pots, Flowerpot Island, Georgian [...] rock of the Flowerpot is Guelph, the protective [...]

CANADA
DEPARTMENT OF MINES
Hon. Martin Burrell, Minister; R. G. McConnell, Deputy Minister

GEOLOGICAL SURVEY
William McInnes, Directing Geologist

MEMOIR 111

No. 91, Geological Series

The Silurian Geology and Faunas of Ontario Peninsula, and Manitoulin and Adjacent Islands

BY

M. Y. Williams

OTTAWA
J. de LABROQUERIE TACHÉ
PRINTER TO THE KING'S MOST EXCELLENT MAJESTY
1919

No. 1703

CONTENTS.

57237—1½

C APTE? VI.

CHAPTER VII.

Illustrations.

The Silurian Geology and Faunas of Ontario Peninsula, and Manitoulin and Adjacent Islands.

CHAPTER I.

INTRODUCTION.

PURPOSE AND CHARACTER OF REPORT.

The "Geology of Canada," by Sir Wm. Logan, published by the Geological Survey of Canada in 1863, includes the only comprehensive account so far published of the Silurian system of southwestern Ontario. This report has long been out of print and so is inaccessible except in the better equipped scientific libraries. Since 1863, many advances have been made in the science of geology, and it is consequently desirable that such fundamental knowledge as that relating to the Silurian rocks of the oldest settled parts of Ontario should be thoroughly revised and brought up to date. Not only is it a part of broad and enlightened education to have a clear knowledge of natural surroundings, but the proper development of our mineral resources and even our agriculture depends upon a clear understanding of geological facts. That no metal mines occur in the area described is true, but the production of non-metallic products such as lime, building stone, salt, gypsum, natural gas, etc., is and has been of the greatest importance in the development of the Ontario peninsula which is one of the most thickly settled parts of Canada.

Taken in conjunction with the recent memoir of this department on the "Devonian of southwestern Ontario", by Clinton R. Stauffer, the present memoir completes the general description and mapping of the geology of southern Ontario as far east as the areas underlain by rocks belonging to the Ordovician system.

As described in the summary reports, the field work upon which this report is based was done during the summers of 1912 to 1916 inclusive. The whole field, including the most inaccessible parts of Manitoulin and adjacent islands and Bruce peninsula, was covered by the author or his assistants; hundreds of miles of formation boundary were followed and surveyed; numerous geological sections were studied and measured; zonal and formational collections of fossils were made, including thousands of individuals, and representing the various geological units throughout their range in southwestern Ontario, economic deposits were studied; and special trips were made to important areas in New York, Michigan, Wisconsin, and Illinois for the purpose of comparing geological occurrences in Ontario with those of the adjacent well-known regions.

ACKNOWLEDGMENTS.

Special acknowledgments are due Professor Charles Schuchert of Yale university, who spent two weeks with the author in the field during the preliminary work of 1912, and one week in 1913. His contributions to the knowledge of the Medina-Cataract problem are acknowledged elsewhere. Professor Schuchert has also kindly criticized this memoir in regard to its more important palæontological and stratigraphical conclusions.

Professor George H. Chadwick of the university of Rochester, guided the author in his studies of the Cayugan formations in western New York state and has assisted materially in solving the Cayugan problems in the Niagara peninsula. His kindness in allowing the author access to his unpublished manuscript on "The stratigraphy of the New York Clinton" has made it possible to bring the Clinton nomenclature of this memoir into harmony with the latest interpretation for New York state. His criticism of this memoir has been most helpful and constructive.

Doctors E. O. Ulrich and R. S. Bassler have kindly identified *bryozoa* and *leperditia* sent them by the author.

The assistance and kindness shown the author by managers of companies and others in the regions visited, have been acknowledged in the annual summary reports, where the names of student assistants are also mentioned. Special recognition, however, is due George S. Hume who assisted the author for two successive years and took charge of the mapping of the Niagara escarpment from Wiarton to Niagara Falls.

The author has been in frequent consultation with E. M. Kindle, palæontologist of the Geological Survey, Canada, during the office and laboratory work required in the preparation of this memoir, and in 1913 Dr. Kindle accompanied him on a reconnaissance trip over the most important sections along the Niagara escarpment between Clinton, New York, and Collingwood, Ontario. His criticism, advice, and assistance are gratefully acknowledged.

The fossil material studied by the author includes, besides his own and his assistants' collections, collections made by E. M. Kindle in preparation of the Niagara Folio, the collections made by J. Stansfield and H. V. Ellsworth, in 1912, and the old collections in the possession of this Survey.

PREVIOUS WORK.

Many eminent geologist have published descriptions of various phases of the geology of so western Ontario, the following being a brief review of the more important contributions to date. For publications not mentioned the reader is referred to the selected list given under Bibliography.

The Rogers brothers of Pennsylvania contest with Alexander Murray, assistant to Sir William Logan, the honour of publishing the first description of the Silurian formations of Ontario. In the report of 1843, the first to be issued by the newly organized Geological Survey of Canada, Murray traced the formations of western New York into western Ontario, then western Upper Canada, giving a general description and outline of their areal extent. A similar report was published the same year by

Rogers and Rogers in the " Transactions of the A erican Philosophical Society ". It devolved upon the Canadian Survey, however, to complete the task of surveying the Ontario region. Murray, in spite of uch other work, returned fro tile to tile to his original task, describing the east side of Bruce peninsula and the Manitoulin islands in the report of 1847-48, the west side of Ontario peninsula fro the north end of Bruce peninsula to Amherstburg, in 1848-49, and rounding out the field work by co pleting the apping of the Niagara escarpment as described in the report of 1850-51. The geological map of western Ontario first appeared as part of a map of the region fro the west end of lake Ontario to lake Michigan, in the Canadian Institute proceedings in 1854, acco panying an article by Sir Willia Logan "On the physical structure of the western district of upper Canada". Meanwhile the solving of geological proble s went on harmoniously in Canada and New York state, the heads of the two Surveys, Sir William Logan and Ja es Hall, respectively, being on the best of ter s. In 1852 Hall published the "Palæontology of New York", vol. II, which greatly facilitated working out the palæontology and stratigraphy of Ontario. In 1863, Logan published his "Geology of Canada", in reality as stated, a su ary of the work of the Canadian Geological Survey fro the beginning. With aps, illustrations of fossils, and descriptions of for ations, this is still the only work that clai s to be a geology of Canada and its excellence is recognized by all authorities. Between the years 1863 and 1910 any publications dealing directly or indirectly with the Silurian syste of Ontario appeared in govern ent reports of Canada and the United States and in various periodicals. T. Sterry Hunt, che ist to the Canadian Geological Survey, published any observations on the geology of Ontario, including notes on the salt and gypsu deposits; Robert Bell described the geology of the Manitoulin islands; Alexander Winchell wrote on the "Geology of petroleu in Canada west"; H. A. Nicholson wrote on the "Palæontology of Ontario"; Hall further described the Niagara and Lower Helderberg for ations of United States and Canada; Whiteaves described the Guelph fauna of Ontario; and Clarke and Ruedemann described the Guelph fauna of New York. These and other publications of like character added infor ation but ade no basic changes in the geology as laid down by Logan in 1863.

In 1910 the "Monroe for ation", by Grab and Sherzer, appeared as a report of the Geological and Biological Survey of Michigan. This vigorously attacked the accepted interpretation of the age of those for ations of western Ontario which had for erly been considered the oldest Devonian for ations of the region. As a result an awakened interest was centred on western Ontario, cul inating for the Devonian system in the "Report on the Devonian of southwestern Ontario", by C. R. Stauffer, and for the Silurian system in papers by Charles Schuchert, Grabau, and the writer, and the present e oir. The eeting of the Twelfth International Geological Congress in Toronto in 1913 necessitated uch geological housecleaning, and when Professor Charles Schuchert acco panied the writer over the Silurian rocks of the Georgian Bay region and Manitoulin island and later went with Dr. W. A. Parks over the Silurian sections of Credit Forks, Ha ilton, and Grimsby, he saw clearly that new correlations for the so-called Clinton and Medina horizons ust be ade: Schuchert's work has been followed up by the writer, along with a revision of the whole Silurian system in southwestern Ontario.

The economic side of the Silurian geology of Ontario has been dealt with in recent reports as follows: "Building and ornamental stones", W. A. Parks, 1912; "Gypsum in Canada", L. H. Cole, 1913; "Salt deposits of Canada", L. H. Cole, 1915; "Oil and gas in Ontario and Quebec", W. Malcolm, 1915; "Records of wells drilled for oil and gas in Ontario", Cyril W. Knight, 1915; "Petroleum and natural gas resources of Canada", F. G. Clapp, 1914-15. (See Bibliography for fuller references.)

BIBLIOGRAPHY.

Bassler, R. S.—"Niagara and Rochester Bryozoa." U.S. Geol. Surv., Bull. No. 292, 31 pls., 1906, p. 136.
"Bibliographic index of American Ordovician and Silurian fossils." U.S. Nat. Mus., Bull. No. 92, vols. I and II, 1915.
Bell, Robert.—"Report on the Manitoulin islands." Geol. Surv., Can., Rept. of Prog., 1863-1866, pp. 165-170.
"Report on the geology of the Grand Manitoulin, Cockburn, Drummond, and St. Josephs islands." Geol. Surv., Can., Rept. of Prog., 1866-1869, pp. 109-116; Map 1870.
"The petroleum fields of Ontario." Trans. Roy. Soc. Can., vol. V, sec. 4, 1888, pp. 101-113.
"The geology of Ontario, with special reference to economic minerals." Rept. of Royal Commission on the mineral resources of Ontario; Toronto, 1889, pp. 57.
"Report on the geology of the French River sheet, Ont." Geol. Surv., Can., new ser., Vol. IX, 1898, pt. I, pp. 29.
Billings, E.—Geol. Surv., Can., "Palaeozoic fossils," vol. I, 1861-65.
Brumell, H. P.—"On geology of natural gas and petroleum in southwestern Ontario." Bull. Geol. Soc. Am., vol. IV, May 20, 1893, pp. 225-240.
"Natural gas and petroleum in Ontario." Geol. Surv., Can., vol. V, pt. Q, pp. 1-94, 5 pls.
Butts, Charles.—"Geology and mineral resources of Jefferson county, Ky." Kentucky Geol. Surv., ser. IV, vol. III, pt. II, 1914-15.
Chadwick, G. H.—"Cayugan waterlimes of western New York." (Abstract) Bull. Geol. Soc. Am., vol. 28, No. 1, p. 173.
"Stratigraphy of the New York Clinton." Bull. Geol. Soc. Am., vol. 29, pp. 327-368.
Chamberlain, T. C.—"Geology of eastern Wisconsin." Geol. of Wisconsin, vol. II, pt. II, 1873-79.
Geology of Wisconsin. vol. I, pt. I, 1873-79.
Chapman, E. J.—"A popular and practical exposition of the minerals and geology of Canada." 1864, p. 190.
"An outline of the geology of Ontario." Can. Jour., vol. XIV, new ser., 1875, pp 580-589.
"An outline of the geology of Canada based on a subdivision of its provinces into natural areas." Pls. 12, vel. XXXIII, pp. 105, Toronto, 1876.
Clapp, F. G.—"Petroleum and natural gas resources of Canada." Dept. of Mines, Mines Branch, vol. II, 1915.
Clarke and Ruedemann.—"Guelph fauna." N.Y. State Mus. Mem. 5, 1903.
"The Eurypterida of New York." New York State Mem. 14, vols. I and II, 1912.
Cole, L. H.—"Gypsum in Canada." Dept. of Mines, Mines Branch, 1913.
"The salt deposits of Canada and the salt industry." Dept. of Mines, Mines Branch, 1915.
Dana, J. D.—"Manual of geology." Revised edition, 1864.
Foerste, A. F. —"Notes on Clinton group fossils with special reference to collections from Indiana, Tennessee, and Georgia." Proc. Nat. Hist. Soc. Boston, vol. XXIV, pls. 5, 1888-9, pp. 263-355.
Twenty-eighth Ann. Rept., Indiana Dept. Geol. and Nat. Res., 1904, pp. 34-35.
"Silurian fossils from the Kokomo, West Union, and Alger horizons of Indiana, Ohio, and Kentucky." Jour. Cincinnati Soc. of Nat. Hist., vol. XXI, No. 1, Sept. 1909, 2 pls.

"Fossils from the Silurian formations of Tennessee, Indiana, and Kentucky." Bull. Denison Univ. Sc. Lab., vol. XIV, 4 pls., 1909, pp. 61-116.

"Upper Ordovician formations in Ontario and Quebec." Geol. Surv., Can., Mem. 83, 1916.

Gibson, J.—"The salt deposits of western Ontario." Am. Jour. Sc., 3d. ser., vol. V, 1873, pp. 362-369.

Goldthwait, J. W.—"An instrumental survey of the shore-lines of the extinct lakes Algonquin and Nipissing in southwestern Ontario." Geol. Surv., Can., Mem. 10, 1910.

Grabau, A. W.—"Guide to geology and palaeontology of Niagara falls and vicinity." Bull. N.Y. State Mus., No. 45, 1901.

"Physical characters and history of some New York formations." Sc., new ser., vol. XXII, 1905, pp. 528-535.

"A revised classification of the North American lower Palaeozoic." Sc., new ser., vol. XXIX, 1909, pp. 351-359.

"Physical and faunal evolution of North America during Ordovician, Siluric, and early Devonic time." Jour. Geol., vol. XVII, No. 3, 1909, pp. 209-252, 11 figs.

"Palaeozoic delta deposits of North Arm." Bull. Geol. Soc. Am., vol. XXIV, No. 3, Sept. 1913, pp. 399-528.

Grabau, A. W., and Sherzer, W. H.—"The Monroe formation of southern Michigan and adjoining regions." Mich. Geol. and Biol. Surv., Pub. No. 2, Geol. Ser. 1, 1910.

Hall, James.—"Palaeontology of New York," vol. II, 1852.

"Account of some new and little known species of fossils from rocks of the age of the Niagara group (in Iowa and Wisconsin)." 12th Ann. Rept., Univ., State of New York, 1867.

"On the relations of the Niagara and lower Helderberg formations, and their geographical distribution in the United States and Canada." Proc. Am. Assoc., vol. XXII, pt. II, 1874, pp. 321-335; Can. Nat., new ser., vol. VII, 1875, pp. 157-159.

Hartnagel, C. A.—"Classification of the geologic formations of the State of New York." New York State Mus. Handbook 19, 1912.

Heinrich, O. J.—"The Manhattan salt mine at Goderich, Canada." Am. Inst. Min. Eng., vol. VI, pl., 1879, pp. 227-274.

Hume, G. S.—"Palaeozoic rocks of lake Timiskaming area." Geol. Surv., Can., Sum. Rept., 1916, pp. 188-192; with map.

Hunt, T. S.—Am. Jour. Sc., 2d. ser., vol. XLVI, 1868, pp. 355-362.

"On the Guelph limestones of North America and their organic remains." Geol. Mag., new ser., vol. II, 1875, pp. 343-348.

"The Goderich salt region." Trans. Am. Inst. Min. Eng., vol. V, 1877, pp. 538-560. Abstract: Am. Jour. Sc., 3d. ser., vol. XIII, 1877, pp. 231-234.

"On the Goderich salt region and Mr. Attrell's exploration." Geol. Surv., Can., Rept. of Prog., 1876-77, pp. 221-243.

Kindle, E. M.—"The stratigraphy and palaeontology of the Niagara of northern Indiana." 28th Ann. Rept., Indiana Dept. Geol. and Nat. Res., 1904, p. 408.

"What does the Medina sandstone of the Niagara section include?" Sc., N.S., vol. XXXIX, No. 1016, June 19, 1914, pp. 915-918.

"Deformation of unconsolidated beds in Nova Scotia and southern Ontario." Bull. Geol. Soc. Am., vol. 28, 1917, pp. 323-334.

Kindle, E. M., and Taylor, F. B.—Niagara folio: U.S. Geol. Folio No. 190, 1913.

Knight, C. W.—"Records of wells drilled for oil and gas in Ontario" 24th Ann. Rept. Ont. Bureau of Mines, pt. II, 1915.

Lambe, L. M.—"A revision of the genera and species of Canadian Palaeozoic corals." Cont. to Can. Pal., vol. IV, pts. I and II, 1899-1900.

Lane, A. C.—Geol. Surv., Michigan: vol. I, 1873.

Logan, W. E.—"On the physical structure of the western district of upper Canada." Proc. Can. Inst., vol. III, pls. 2, 1854, pp. 1-2.

Geol. Surv., Can., Rept. of Prog., from its commencement to 1863.

McLearn, F. H.—"The Silurian Arisaig series or Arisaig, Nova Scotia." Am. Jour. Sc., vol. XLV, 1918.

Malcolm, Wyatt.—"Oil and gas fields of Ontario and Quebec." Geol. Surv., Can. Mem. 81, 1915.

Miller, W. G.—"The limestones of Ontario." Rept. Ont. Bureau of Mines, pt. II, 1904.

Murray, A.—"Report on district lying between Georgian bay, on lake Huron, and the lower extremity of lake Erie." Geol. Surv., Can., Rept. of g., 1843.

"On north coast of lake Huron." Geol. Surv., Can., Rept. of Prog., 1847-48, 1849, pp. 94-106.

6

"On Bruce peninsula and Manitoulin islands." Ibid, pp. 118-121.

"On further examination of the shores, islands, and rivers of lake Huron." Geo Surv., Can., Rept. of Prog., 1848-49, pp. 7-46.

"On work in the great peninsula bounded by lakes Huron, St. Clair, and Erie. Geol. Surv., Can., Rept. of Prog., 1850-51, pp. 13-33.

Nattress, Thos.—"The contour of the Sylvania sandrock and related strata of th Detroit River area." 12th Rept. Mich. Acad. Sc., 1910.

"The extent of the Anderdon beds of Essex county, Ontario, and their place in th geologic column." 13th Rept. Mich. Acad. Sc., 1911.

"Additional notes on the geology of the Detroit River area." 14th Rept. Mich Acad. Sc., 1912.

"Geology of the Detroit River area." Rept. Ont. Bureau of Mines, 1912, pp. 28, 287

Nicholson, H. A.—"Palæontology of the province of Ontario," 1874;

"Guelph limestone of North America." Geol. Mag., new ser., vol. II, 1875, pp 343-348.

O'Connell, Marjorie.—"The habitat of the eurypterida", Bull. Buffalo Soc. of Nat Sc., 1916.

Parks, W. A.—"Fossiliferous rocks of southwestern Ontario." Rept. Ont. Bureau o Mines, 1903, pp. 141-156;

"Building and ornamental stones of Canada," vol. I. Dept. of Mines, Mines Branch, 1912.

"The Palæozoic section at Hamilton." Twelfth Inter. Geol. Cong., Guide Book No. 4, 1913, pp. 125-142; map and illustrations.

Reinecke, Leopold.—"Road material surveys in 1914." Geol. Surv., Can., Mem. 85, 1916, 244 pp., 10 pls., 2 figs., 5 maps.

Rogers, H. D., and Rogers, W. B.—"Conservatio ... the geology of the western peninsula of upper Canada and the western part of Ohio," Trans. Am. Phil. Soc., new ser., vol. VIII, 1843, pp. 273-284; Proc. Am. Phil. Soc., vol. II, 1842, pp. 120-125.

Ruedemann, Rudolf.—"Note on habitat of the eurypterids." New York State Mus., Bull. 189, pp. 113-115.

Rominger, C.—"Fossil corals." Mich. Geol. Surv., vol. III, pt. II, pls. 55, 1876, pp. 161.

Savage, T. E.—"Stratigraphy and palæontology of the Alexandrian series in Illinois and Missouri." Illinois Geol. Surv., Bull. 23, pls. 9, 1917, pp. 67-160.

"Correlation of the early Silurian rocks in the Hudson Bay region." Jour. Geol., vol. XXVI, 1918, pp. 334-340.

Savage, T. E., and Crooks, H. F.—"Early Silurian rocks of the northern peninsula of Michigan." Am. Jour. Sc., vol. XLV, Jan. 1918, pp. 59-64.

Schuchert, C.—"Palæogeography of North America." Bull. Geol. Soc. Am., vol. XX, pls. 46-101, 1910, pp. 427-606.

"The Cataract: a new formation at the base of the Siluric in Ontario and New York." Bull. Geol. Soc. Am., vol. XXIV, 1913, p. 107.

"Medina and Cataract formations of the Siluric of New York and Ontario." Bull. Geol. Soc. Am., vol. XXV, pls. 13-14, 1914, pp. 277-320.

Sherzer, W. H., and Grabau, A. W.—"New upper Siluric fauna from southern Michigan." Bull. Geol. Soc. Am., vol. XIX, fig. 1, 1909, pp. 540-553.

Stauffer, C. R.—"The Devonian of southwestern Ontario." Geol. Surv., Can., Mem. 34, 1915.

Bull. Geol. Soc. Am., vol. XXVII, 1915, p. 77.

Taylor, F. B.—"The moraine systems of southwestern Ontario." Trans. Can. Inst., 1913, pp. 1-23; map.

Twenhofel, W. H.—"The Anticosti Island faunas." Geol. Surv., Can., Mus. Bull. No. 3, 1914.

Twenhofel, W. H., and Schuchert, C.—"The Silurian section at Arisaig, Nova Scotia, with a correlation note by Charles Schuchert." Am. Jour. Sc., 4th ser., vol. XXVIII, 1909, pp. 143-164.

Ulrich, E. O.—"Revision of the Palæozoic systems." Bull. Geol. Soc. Am., No. 3, vol. XXII, pls. 5, Sept. 1911, pp. 281-680.

"The Medina problem (abstract) Bull. Geol. Soc. Am., No. 3, vol. XXIV, pp. 107-108, 1913.

"The Ordovician-Silurian boundary." Twelfth Inter. Geol. Cong., Can., 1913, C.R., pp. 593-667, maps 8, table 1.

Whiteaves, J. F.—"On some new and imperfectly characterized or previously unrecorded species of fossils from the Guelph formation of Ontario." Palæozoic fossils, Geol. Surv., Can., vol. III, pt. I, 1895.

" Revision of the fauna of the Guelph formation of Ontario with descriptions of a few new species." Palæozoic fossils, Geol. Surv., Can., vol. III, pt. IV; 1906.
Whitfield, R. I " Silurian palæontology." Geol. of Wis., vol. IV, 1882, pp. 313-319.
Williams, M. "The Silurian of Manitoulin island and western Ontario." Geol. Surv., Can. Sum. Rept., 1912, pp. 275-284.
" Revision of the Silurian of southwestern Ontario." Ottawa Nat., vol. XXVII, 1913, pp. 3, 38.
" Stratigraph
Can., Sum. Rept., 1913, pp. 178-188.
" Arisaig-Antigonish district, Nova Scotia." Geol. Surv., Can., Mem. 60, 1914.
" The middle and upper Silurian of southwestern Ontario." Geol. Surv., Can., Sum. Rept., 1914, p. 82-86.
" The Ordovician rocks of lake Timiskaming." Geol. Surv., Can., Mus. Bull. No. 17, 1915.
" An eurypterid horizon in the Niagara formation of Ontario." Geol. Surv., Can., Mus. Bull. No. 20, 1915.
" Formations adjacent to the Niagara escarpment of southwestern Ontario." Geol. Surv., Can., Sum. Rept., 1915, pp. 139-142.
" Investigations in Ontario." Geol. Surv., Can., Sum. Rept., 1916, pp. 186-188.
W   -" index to the stratigraphy of North America." U.S. Geol. Surv., Prof. 71, 1912, p. 25-268; geol. map of North America.
W. G —" The theory of the formation of sedimentary deposits." Can. Rec. d. IX. No. 2, 18-4, pp. 112-132.
W   A.— Note on the geology of petroleum in Canada west." Am. Jour. Sc., , vol. XLI, 18 pp. 176-178.

CHAPTER II.

GENERAL CHARACTER AND GEOLOGY OF SOUTHWESTERN ONTARIO.

INDUSTRIES.

That part of southwestern Ontario underlain by rocks of Silurian age, includes some of the richest and most thickly settled agricultural districts in Canada. With splendid transportation facilities, manufacturing industries have developed rapidly in centres such as Welland, Brantford, Guelph, Kitchener, Stratford, and a host of smaller places. The shallow soil near the edge of the Niagara escarpment is largely covered by wood lots or is used for pasture. The rocky land on Bruce peninsula and Manitoulin island, now that the best of the timber has been harvested, is utilized for farming and grazing in the more suitable areas, and elsewhere is left for the growth of hardwood. Excellent farm land occurs in parts of Bruce peninsula, particularly west of Lyon head, and much good farm land is worked on Manitoulin island, areas near Manitowaning, Little Current, and Mindemoya, at the head of South bay, Gore bay, and on Barrie island, being settled by prosperous farming communities.

Throughout the Georgian Bay and Lake Huron regions, fishing is an important industry and the abundance of game fish, combined with the natural picturesqueness and variety of scenery, are strong attractions for summer tourists.

The mineral products obtained from the Silurian rocks consist of commonplace non-metallics, such as road metal, building stone, quicklime, salt, gypsum, gas, and some oil. These products, however, are of great potential value when their proximity to the centres of population is taken into account. The splendidly ordered gypsum mine at Caledonia formerly owned by the Alabastine Company and now owned by the Ontario Gypsum Company, is a revelation to the average man, who disconnects mining entirely with this part of Ontario. The salt wells of the Kincardine, Goderich, and Sarnia regions practically control the Canadian market for salt and in a sense reserves are as yet untouched. The salt production for 1917 is valued at $1,095,866. The gas produced in Ontario in 1917 was valued at $3,182,154 and this was mostly derived from the Silurian formations.

TOPOGRAPHY.

The topography of the Silurian areas is generally of low relief, the only feature of important dimensions being the Niagara escarpment which forms a step upward to the west and south from the lowlying Ordovician plains to the east and north. This sinuous escarpment is the outcropping edge of the resistant Niagara (Lockport) dolomite, into which the agents of erosion have eaten. On the mainland of Ontario, the eastern extension of the Silurian system is practically terminated by the Niagara escarpment; but the Manitoulin dolomite and the Whirlpool sandstone, at the base of the Cataract formation, extend in places considerable distances to the east or north from beneath the Niagara dolomite, forming at their eroded edges one or more additional, but insignificant escarpments. On Manitoulin island the escarpment formed by the Manitoulin dolomite is comparatively conspicuous. To the west and south of the Niagara escarpment, the land surface in general tends to slope away to the southwest down the general dip of the formations. Numerous modifications occur, however, due to accumulations of glacial and outwash materials and to glacial and other erosion of the hard rock surfaces. Examples of eroded rock surfaces are common along the Niagara escarpment, on Bruce peninsula, and on Manitoulin and the adjacent islands. On Bruce peninsula and the islands to the north, the solid dolomite has been sculptured into innumerable rounds, hills, and valleys, the latter containing more or less gravel, clay, and soil. Excepting the southern half of Niagara peninsula which is mostly very level, the areas underlain by Silurian rocks are generally undulatory, the more prominent hills being due for the most part, as at Durham, to the presence of glacial moraines, and the more pronounced valleys being due to river channels, such as those of the Saugeen and the Grand. Where the drift is deep, the outcrops of rock are practically confined to the river valleys.

The maximum recorded elevation for the region described is 1,700 feet A. M. T. at Dundalk. The maximum for the Manitoulin island is 1,020 feet A. M. T. southeast of lake Manitou. The minimum elevation for any part of the country obviously depends upon the elevation of the great lake into which its streams empty.

GENERAL GEOLOGY.

Southwestern Ontario lies in the northern, central part of the Palæozoic sedimentary area which occupies the basin of the Great Lakes, exclusive of lake Superior, extends from the Appalachian mountains on the east to Kansas and Nebraska on the west, and includes the drainage basin of the Mississippi river as far south as Oklahoma, northern Tennessee, and central Alabama. Of this vast area, the rocks of Silurian age outcrop over only a small part. The other systems of Palæozoic rocks represented in Ontario are the Cambrian, Ordovician, and Devonian. The Carboniferous system with its coal beds is found nowhere within the province, but outcrops to the west in Michigan and to the south in Ohio. The Palæozoic formations rest upon an uneven floor of Pre-Cambrian, igneous and metamorphic rocks, which consist of granites, diabases, gabbros, gneisses, schists, quartzites, crystalline limestones, etc. This basal complex outcrops to the north as the Canadian Pre-Cambrian shield. Corresponding to the gentle southerly dip of the Pre-Cambrian floor, the sedimentary strata in southwestern Ontario are inclined in general to the southwest with an average dip of about 20 feet a mile. East of lake St. Clair, however, a basin structure occurs and there is in consequence an easterly dip in the region south of lake St. Clair and east of Detroit river. The maximum thickness of the Palæozoic strata, as taken from a well record in the basin just mentioned, is 3,635 feet, of which 35 feet is basal arkose or conglomerate, 1,625 feet includes the Ordovician and any lower Palæozoic strata that may be present, 1,725 feet is the thickness of the Silurian system, and 280 feet is the thickness of the Devonian formations present.

57237—2

10

On the east and north the Silurian rocks are bounded approximately by the Niagara escarpment, which is a well-defined and conspicuous feature of the landscape from central New York to Niagara Falls, and thence through Hamilton, to the vicinity of Collagwood, along the south and

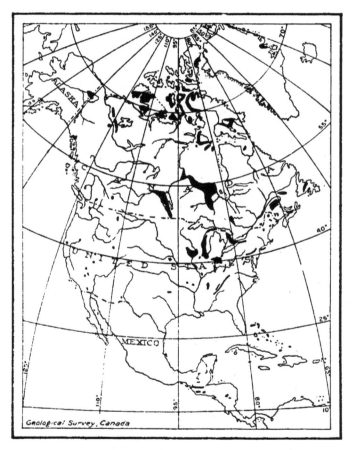

Figure 1. Diagram showing, in black, the distribution of Silurian rocks in North America.

western shores of Georgian bay, through a chain of islands to Manitoulin island, throughout the length of Manitoulin, Cockburn, and Drummond islands to the northern peninsula of Michigan. The islands separating Georgian bay and the North channel from lake Huron are so continuous and so closely connected by reefs that the break here is more apparent than real.

Although Silurian formations underlie the whole area from the Niagara escarpment to the limits of the province on the south and west, the area not overlain by Devonian strata is much more restricted, as may be seen from the map, and the actual exposures of rock occupy only a small part of this area.

The sandstones, shales, and interbedded dolomites, comprising the lower part of the Silurian system (the Medina-Cataract, the Clinton, and Rochester subdivisions) outcrop along the front of the Niagara escarpment, or in case of the harder subdivisions, occur by themselves to the east or north as more or less localized escarpments. The overlying, heavy-bedded dolomite of Niagara age (Lockport) forms the practically continuous white cliffs of the Niagara escarpment.

West and south of the Niagara escarpment, the Lockport dolomite outcrops over wide areas of country, particularly on Bruce peninsula and Manitoulin and the neighbouring islands, where large areas of barren rock are exposed. The Guelph dolomite overlies the Niagara (Lockport) dolomite and outcrops over considerable areas to the south and west of the Niagara belt, continuing as far north as Fitzwilliam island. The waterlime, shales, and gypsum beds of the Cayugan group are rarely seen at the surface, sections occurring as indicated on the map in the valleys of Grand River, Saugeen river, and in some other stream channels. The salt beds do not outcrop at all.

In many parts of Ontario peninsula, bedrock is deeply covered by surface deposits, consisting of boulder till, and outwash materials such as clay, gravel, sand, etc. The study of this surficial geology is an extensive subject in itself and has been described in part by J. W. Goldthwait and F. B. Taylor (see Bibliography).

In areas where rock exposures are scarce owing to the thickness of the surface deposits, the land is generally good and the country is thickly populated. In consequence, rock outcrops are of great potential value as a source of raw material for building stone, road metal, lime manufacture, etc., and those discovered in the preparation of this report have been indicated on the map.

ROCK STRUCTURE.

Since Pre-Cambrian time, southwestern Ontario has not suffered in any way from violent deformation, but on the other hand the movements of the earth recorded elsewhere in faults and sharp folds have resulted in gentle structural undulations, which rarely assume a magnitude of dip of 200 feet per mile, but which are nevertheless controlling factors in the accumulation of oil and gas. The oil fields of Essex, Kent, Lambton, and Middlesex counties are located for the most part on gentle domes and anticlines, and the gas fields between Niagara and Detroit rivers, where studied, have been shown to be closely associated with inverted basin structures, although varying rock porosity in some cases is the primary controlling factor of gas as well as of oil occurrence, and in formations that do not carry water the oil and gas may be found in the basins and synclines.

A few conspicuous indications of structure have been cited in previous reports. In the "Geology of Canada", 1863, and again as restated by Malcolm[1] the Cincinnati anticline which is a decided feature in the Ohio valley, "forms a gentle curve running from the west end of lake Ontario, through Woodstock, along the Thames, and through Chatham to the southern part of Essex county. This diffuse fold is possibly composed of subordinate folds, none of which are pronounced or have limbs dipping at more than a few feet per mile." The diffuseness of this fold is well demonstrated by a detailed study made by the author of the structure at the Bothwell oil field where low minor folds are quite complex in arrangement. There are very good grounds for locating the broad anticline as above stated, but it is well to emphasize the diffuseness of its characters and not to expect a continuous, clearly defined, structural feature. It is probably a zone of low folds rather than a single well-marked anticline.

Anticlines have been located by Bell on the map of Manitoulin island, and judging from the shape of the topographic features of the land some of these at least may be as indicated. Much more detailed work is required, however, to interpret satisfactorily the comparatively gentle structure of Manitoulin island.

Two anticlines at Rockwood were described by Logan.[2] The one at the top of the Niagara (Lockport) is a well-marked structure and is further described below. The reference to a cross fold from Eden through Rockwood to Orangeville, appears to be based upon re-entrants in the supposed boundary between the Guelph and Lockport formations, re-entrants which do not exist in the revised boundary. The marked anticline in the top of the upper Lockport dolomite, which extends through Rockwood village, is very clearly defined 200 yards southwest of the road bridge crossing the Eramosa river. Here, the river cuts through the crest of the anticline which for a width of about 100 yards has dips of about 5 degrees to the east and west. Farther from the crest the dip is 10 to 15 degrees on the east and somewhat less on the west. Eastward as far as the river and south of the railway track, the strata dip at the rate of 4 to 5 degrees to the east, but for three-quarters of a mile north of the railway bridge, the river bed is on the crest of a low anticline which merges into the more pronounced structure south of the railway. A small syncline intervenes between this anticline and the crest of the main anticline west of Rockwood station, and eastward from the railway bridge a similar syncline occurs, succeeded eastward by a low anticline. Thus it is clear that the main anticline is flanked to the eastward by minor folds. To the west the dip is clearly shown, although undulatory, as far as a small remnant of Guelph dolomite south of the Grand Trunk railway, about 400 yards west of Rockwood station. The Eramosa beds which form the upper part of the Lockport formation intervene between the rock at the village and the Guelph remnant to the west and allowing

[1] Geol. Surv., Can., Mem. 81, p. 45.
[2] "Geology of Canada", 1863, pp. 330 and 343.

for their thickness (40 to 50 feet) and averaging the rock dips, it is estimated that the dip to the west is about 70 feet in 1,200 feet, or at the rate
of about 300 feet per mile. As the average dip of the formation to the
west is about 20 to 30 feet per mile, it is clear that the anticline is well
marked. The crest of the anticline extends through the Agricultural
grounds, the west side of the schoolyard, and through the middle of the
rock-cut on the railway about 235 yards west of the station, and crosses
the road between concessions IV and V, Eramosa township, near the
north side of lot 6. The axis of the anticline extends nearly northwest
from Eramosa river, but curves more and more to the north in its northern
exposures. To the south of Rockwood the rock is covered with a thick
mantle of surface material.

Stauffer[1] described an anticline passing through St. Marys as follows:
" Running nearly north and south through the city there is a rather pronounced anticlinal or monoclinal fold............which brings up the
Silurian[2] to the east and drops the Devonian to the west. The Thames
river cuts into the side of this fold so that at the dam near the Queens
Street bridge the dip is upstream, while a quarter of a mile below the
bridge, the dip is downstream. The excellent artesian wells which supply
St. Marys with such a quantity of good water may depend on this
same structure."

Folds have been mentioned at other localities, but it is difficult to
prove their existence.

As may be seen from the cross section passing through Elora (see Map
1715), there evidently is a monoclinal fold to the west of that town. Such
a fold is strongly suggested by comparing the altitude of the top of the
Guelph outcrops at Elora and the altitude of the top of the formation
derived from the depth at which Guelph dolomite was struck in a well
drilled on lot 6, concession V, Pilkington township, about 5 miles west of
Elora.

Undulations are clearly marked along the north end of Bruce peninsula
between Cabot head and Tobermory, a clearly cut escarpment being
exposed all the way and the water-level being everywhere a handy reference
plane. The best marked anticline extends from the outlet of Gillies lake
northwesterly to the small area of Cataract exposures on the northern
shore. The easterly dip of this fold as seen on the north shore is about
110 feet in a mile, the corresponding dip at Gillies lake being much less.
The dips to the west are similar. A well-marked syncline parallels the
anticline on the east and includes the three conspicuous bluffs west of
Cabot head, whose axes are at right angles to that of the syncline. Farther
west the undulations are not so easily determined because of the lack of
horizon markers, but they are quite evident to an observer coasting along
the shore. Other slight undulations may be seen along the east side of
Bruce peninsula.

The gentle monoclinal structure of the Niagara peninsula is clearly
illustrated by the three structural sections accompanying the special map
of this region. Kindle states[3] that "Between the New York Central Railway bridge and the elevator shaft at the falls the beds dip to the south at

[1]Geol. Surv., Can., Mem. 34, p. 113.
[2]Detroit River series and hence now considered Devonian.
[3]U.S. Geol. Surv., Niagara Folio No. 190, 1913, p. 14.

an average rate of 28½ feet to the mile. The rate of dip along the northe two-thirds of the gorge is somewhat less...........................

"A group of more than twenty wells drilled for the Tonawanda Ga Company affords complete data concerning the rate of dip............ about Getzville....................The available facts indicate tha along an east-west belt of territory..............the southerly dip (the rocks is about 10 or 12 feet greater to the mile than in the contiguou territory on the north".

Getzville is 5 miles east of Tonawanda and almost due east of th comparatively steep dips indicated in the section passing through Niagar; falls. At Getzville, this change of dip has caused the accumulation o an important gas pool in the Medina sandstones.

To the westward the monoclinal structure merges into the broad diffuse east-west fold of the Cincinnati anticline which is generally con sidered as having its axis along the Dundas valley[1].

[1]Geol. Surv., Can., vol. V, pt. Q., pl. V, pp. 1-94.

CHAPTER III.

SUMMARY AND CONCLUSIONS.

GENERAL GEOLOGY.

The Silurian system is bounded below, in Ontario peninsula, by the red Queenston shale and on Manitoulin island by the Richmond shales and limestones. These underlying formations belong to the Cincinnatian group of the Ordovician system and are only facies of the same sedimentation. Above, the Silurian is bounded by different Devonian formations, the Oriskany sandstone and Onondaga limestone in the Niagara peninsula, and the Sylvania sandstone and Detroit River series in the Lake Huron, Lake St. Clair, and Detroit River regions. In intermediate areas the facts are less definitely known. The total thickness of Silurian strata at lake St. Clair is from 1,725 to 1,750 feet.

The Silurian system falls lithologically into three subdivisions, a lower one typically marked by an alternation of clastic and calcareous sediments, the result of changing conditions of the land in respect to the sea; a middle group of massive dolomites, suggestive of widespread seas of moderate depth; and an upper division of saline sediments containing lenses of salt and gypsum and impure, clastic dolomites. The saline deposits are probably those of shallow, interior basin, water bodies, which received great quantities of detrital material from the surrounding areas where older formations elevated above sea-level were being actively eroded. The last phase of Silurian sedimentation represented a brief return to normal marine conditions, which were terminated by the uplift preceding Devonian time.

The oldest Silurian division includes: (1) the sandstones and shales of the Niagara peninsula (Medina) and the contemporaneous dolomites, shales, and shaly dolomites (Cataract) of the Georgian Bay region; (2) the limestones of the Clinton formation which scarcely extend in outcrop beyond the Niagara peninsula; and (3) the Rochester shale which thins out a few miles north of Hamilton.

The middle division includes the widespread Niagara dolomite, suitably known as Lockport dolomite; and the Guelph dolomite which is indicative of a shallowing sea, approaching interior basin conditions.

The upper division of the Silurian includes the Cayugan group, with its Salina shales, gypsum and salt, these being the extension of the Camillus shale of New York, and the overlying Bertie waterlime and Akron dolomite, which merge laterally into the lower Monroe dolomites of the Lake Huron-Detroit River region.

The best marked disconformities or breaks in sedimentation occur: (1) at the base of the Silurian where the Whirlpool sandstone and Manitoulin dolomite overlap the Queenston shale and marine Richmond formation; (2) at the base of the Lockport dolomite which successively overlaps from the north the Cataract, the Clinton, and the Rochester formations; (3) at the base of the Salina which overlies what appears to be an erosion surface at the top of the Guelph formation; and (4) at the top of the Silurian as represented by the Bass Island series in the west and the Akron dolomite in the east. The contact between the Akron dolomite

and the overlying Oriskany sandstone and Onondaga li estone is clear
characterized by an erosion unconfor ity, and the corresponding conta
in the west between the Bass Island series and the Sylvania sandstone
probably si ilar, but it is known in Ontario only fro well records.

ECONOMIC GEOLOGY·

All the Silurian calcareous sedi ents in Ontario are highly agnesia
and are classed as dolo ites. This i plies that where sufficiently fr
fro i i purities such as clay and silica the Silurian dolo ites may be buri
for li e anufacture and are especially suitable for hydrated li e; bu
they are entirely unsuitable for the manufacture of Portland cemen
Guelph dolo ite at Guelph, Fergus, and elsewhere is extensively used f
the anufacture of quickli e. The Manitoulin, Clinton, Lockpor
Bertie, and Akron dolo ites are generally suitable for quicklime manu
facture and have been used to so e extent. Natural rock cement w
for erly ade fro the argillaceous dolomites or "waterlimes" at tl
base of the Lockport dolo ite at Thorold, Mt. Albion, and elsewher
and fro the Bertie "waterlime" of Bertie township. This industry
now abandoned, because of the inferiority of natural rock ce ent to th
high grade Portland ce ent now used in construction work.

The best known Silurian building stones are obtain fro tl
Whirlpool sandstone, the Credit Valley "Brown stone" being ry favou
ably known, and fro the Lockport and Guelph dolo ites. The grea
extent of the exposures of the last two na ed for ations akes ther
easily available for any cities and towns, and it is only because ston
has given place to cheaper building aterials that this fine natural resourc
has been very little developed. The Manitoulin or lower Cataract dolo it
of Manitoulin island is in any cases suitable for building stone, althougl
here as farther south it is co only too thin-bedded and shaly.

The best Silurian rock for road etal includes the upper and harde
beds of the Lockport dolo ite, the denser and harder beds of the Guelpl
dolo ite, and uch of the Akron dolo ite (including probably a part o
the Bertie beds) outcropping near Cayuga, Dunnville, and elsewhere ii
the Niagara peninsula.

Large gypsum deposits occur in the Salina for ation along the Gran
river between Paris and Cayuga. These are at present ined at Caledoni
and York.

Extensive and as yet only partly developed beds of salt occur in th
Salina for ation over a large area extending east fro lakes Huron an
St. Clair, between Kincardine and Port La bton, as far as Brussels
London, and Dutton. Another area is situated in the vicinity of Windsor
These areas are the centre of the salt industry in Canada.

Much gas is derived fro the Medina sandstone (the Whirlpool an
overlying beds) in Niagara peninsula and as far west as Delhi. Th
Clinton li estone furnishes gas in the Onondaga, Caledonia, Cayuga, an
Port Colborne fields. The Guelph dolo ite is gas-bearing in easter
Essex county.

So e oil is obtained fro the Whirlpool sandstone near Brantford (thi
is variously reported fro the Whirlpool and the underlying ueenston
and the Guelph and Salina furnish the oil of the Tilbury field of Ken
and Essex counties.

the
iod,
lose.
ions
ding
ime-
ying

pper
ary.

ian-
nond
gical
gists,
pted
s the
amp-
Rich-
lying
n in
. In
phase
nond.
shale
nized
ever,
t the
mond
tings
break
lpool
n but
beds.
over-
cided
d by

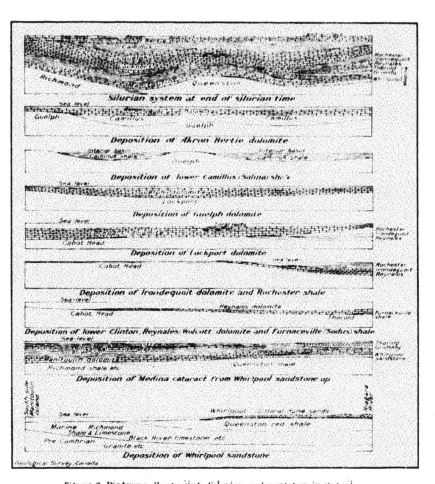

Figure 2. Diagrams illustrating Silurian sedimentation in Ontario.

To accompany Memoir by M Y Williams

CHAPTER IV.

THE SILURIAN SYSTEM.

GENERAL STATEMENT OF EXTENT.

The Silurian system includes the sediments deposited between the emergence of the land from the sea which closed the Ordovician period, and the like emergence which brought the Silurian period to a close. According to the author's interpretation for Ontario, the Silurian formations would be those lying between the sediments of Richmond age (including Queenston shale) below and the Oriskany sandstone, Onondaga limestone, or Sylvania sandstone, whichever happened to be the local overlying formation of the Devonian system, above.

As geologists are not unanimous regarding either the lower or the upper Silurian limits some explanation of the author's conclusions is necessary. Let us first consider the lower Silurian boundary.

LOWER BOUNDARY.

Until recent years, geologists have unanimously placed the Ordovician-Silurian boundary at the top of the well-recognized marine Richmond formation. This conclusion was based on a mass of palaeontological evidence and was in accord with stratigraphic evidence as well. Geologists, following the terminology of New York state, also unanimously accepted the base of the "Red Medina," now known as Queenston shale, as the lower boundary of the Silurian. This decision was based on the assumption that the Queenston shale overlay and was younger than the Richmond formation, and was .. continuous deposition with the overlying "Medina" used in the restricted sense. As no fossils were known in the Queenston shale, palaeontological evidence was entirely lacking. In 1909[1] Grabau showed that the Queenston shale is merely a delta phase of Richmond sedimentation, and hence is of the same age as the Richmond. The majority of geologists at once accepted the top of the Queenston shale as the top of the Ordovician system in accordance with the long recognized treatment of the Richmond marine sediments. A few geologists, however, continued to interpret the Ordovician-Silurian boundary as being at the base of the Queenston and so placed it at the base of the typical Richmond formation. For this view of the case the reader is referred to the writings of E. O. Ulrich.[2] So far as Ontario is concerned a stratigraphic break between the top of the Queenston shale and the overlying Whirlpool sandstone is indicated not only by a decided change in sedimentation but also by mud cracks in the Queenston filled with sand from the Whirlpool beds. In the Georgian Bay region, however, where the Manitoulin dolomite overlies shale of the marine Richmond formation, there is a far more decided break in both sedimentation and fauna. Not a single species listed by

---

[1] Science, new ser., vol. XXIX, p. 356.
[2] "The Ordovician-Silurian boundary". Twelfth Inter. Geol. Cong., 1913. C. R., pp. 593-667.

Foerste from the Richmond of this region is known in the Manito
dolomite, and the genera that cross over are few in number and of l
ranging characters. These facts have caused the author to adopt
more widely accepted view of stratigraphers.

UPPER BOUNDARY.

The problem of the upper contact of the Silurian system in the Det
river-Lake Huron region is even more complicated and obscure t
is that of the lower contact just described. In the Niagara penin
and vicinity the Oriskany sandstone, or in its absence the overlying On
daga limestone (both of undoubted Devonian age) disconformably ove
the Akron dolomite of undoubted Silurian age and the boundary is t
definitely located. From the vicinity of Woodstock westward, howe
the Monroe formation of Grabau intervenes between the easily-recogni
Silurian and Devonian formations, and Grabau has placed all of
series in the Silurian system. Unfortunately for the solution of the pr
lem, outcrops of Monroe strata are scarce, few good sections are kno
and much has to be inferred from samples taken from drill holes.

The accompanying table gives Grabau's interpretation of the Mon
formation. This succession was determined from the studies made
the Michigan geologists during the sinking of a salt shaft near Detr
and from natural sections and bore-holes in Ohio and Michigan. N
withstanding the criticisms of this interpretation of the sequence of
formations by the Rev. Thomas Nattress,[1] there is good reason to belle
that it is substantially correct.

	Dundee formation			
			Disconformity	
	C. Upper Monroe	Lucas dolomite		200 f
	or	Amherstburg dolomite		20
	Detroit River	Anderdon limestone		35- 50
	series	Flat Rock dolomite		40-100
			Disconformity	
	B. Sylvania sandstone and dolomite			30-300
			(Overlap)	
	A. Lower Monroe	Raisin River dolomite..		200
	or	Put-in-Bay dolomite		100
	Bass Island	Tymochtee shales		90
	series	Greenfield dolomite.....		100
			Disconformity	
	Salina formation			

The lower Monroe or Bass Island series is undoubtedly Silurian a
will be described under "Cayugan group." Grabau has demonstrated th
the typical Sylvania sandstone of Michigan is in part of sand dune ori
and hence represents continental sedimentation transgressing the Silur
land. A further discussion of the characters and age of the Sylva

[1] Nattress, Thomas. See Bibliography.

t,
of
st
l-
l-
te
ts
lt
ts

ls
'h
h
e.
s.
m

te
is
n'
m
ng
s-
id
'er
in
za
ns
ne
of
)n
us
id,
ef
ra
ies
to

Boundaries and Subdivisions of the Silurian System of Southwestern Ontario.

System.	Formation.	Localities.			
		Niagara river.	Dundas.	Bruce peninsula.	Detroit river.
Devonian.	Onondaga / Oriskany / Detroit river. / Sylvania.	Onondaga limestone. / Oriskany sandstone and conglomerate in lenses.			Onondaga (Dundee) limestone. / Oriskany? sandstone (thin lenses, age uncertain.) / Detroit River limestone and dolomite, four members. / Sylvania sandstone.
Silurian.	Cayugan group — Akron.	Akron dolomite.			Base Island: Raisin River dolomite. / Put-in-Bay dolomite. / Tymochtee shale. / Greenfield dolomite.
	Cayugan group — Bertie.	Bertie waterlime.			
	Cayugan group — Salina.	Camillus shale including gypsum.	Present land surface.	Present land surface.	Camillus shale including salt and gypsum.
	Guelph.	Guelph dolomite.	Guelph dolomite.	Guelph dolomite.	Guelph dolomite.
	Niagara group — Lockport.	Eramosa beds. / Dolomite. / Gasport dolomitic limestone. / DeCew waterlime.	Eramosa beds. / Dolomite. / Chert beds.	Eramosa beds. / Dolomite.	Lockport dolomite.
	Niagara group — Rochester.	Rochester shale.	Rochester shale (very thin.)		Rochester shale.
	Clinton.	Irondequoit dolomite. / Williamson? shale. / Reynales (Wolcott) dolomite. / Furnaceville? (Sodus) shale.	Irondequoit dolomite (very thin.) / Reynales (Wolcott) dolomite.		Dolomite.
	Medina-Cataract.	Thorold sandstone. / Grimsby sandstone. / Cabot Head shale. / Manitoulin beds (shale and calcareous sandstone). / Whirlpool sandstone.	Grimsby sandstone. / Cabot Head shale. / Manitoulin dolomite. / Whirlpool sandstone.	Cabot St. Edmund Head dolomite. shale Shale. / Dyer Bay dolomite. / Shale. / Manitoulin dolomite.	Cabot Head shale. / Manitoulin dolomite.
Ordovician.	Richmond and Queenston.	Queenston shale.	Queenston shale.	Queenston shale.	Queenston shale.

57237

F(
dc
ra
m

riy
is
an
da
th
de
th
Sil
sei
lei
an

foi
th
an
wi
foi
th

=

—

—

wil
the
an
lan
—

sandstone is to be found below. The controversy, however, centres about the age of the upper Monroe or Detroit River series. The palaeontologic and stratigraphic evidence apparently conflict, and it is, therefore, necessary briefly to review the evidence in order to make the author's conclusions clear.

The Devonian characters of the Detroit River series are well represented by the following fossils: *Syringopora* cf *hisingeri* Billings, in the Flat Rock dolomite; *Cystiphyllum americanum* mut. *anderdonense* Grabau, *Favosites basaltica* mut. *nana* Grabau, *Conocardium monroicum* Grabau (closely related to *C. trigonale* Hall), in the Anderdon limestone; *Romingeria umbellifera* Billings, *Panenka canadensis* Whiteaves, *Conocardium monroicum* Grabau, *Proetus crassimarginatus* Hall, in the Amherstburg dolomite; and *Romingeria umbellifera* (Billings), *Conocardium monroicum* Grabau, *Panenka canadensis* Whiteaves in the Lucas dolomite. Grabau[1] says of the fauna of the Flat Rock, Anderdon, and Amherstburg beds: " Its most characteristic feature is its Devonic element. If the fauna were considered by itself, it would probably be pronounced a Schoharie or an Onondaga fauna without hesitation, though there is a considerable Siluric element. The position of this fauna beneath 200 to 250 feet of the Lucas dolomite with a Siluric fauna, forces us to consider this as Siluric." It will be seen below just how important the Siluric element in the Lucas dolomite is. Continuing, Grabau says[2] " To sum up, the stromatoporoids of this fauna are partly Siluric and partly of Devonic types. The corals are represented by nine Siluric species and thirteen species identified with or most nearly like mid-Devonic species. The brachiopods are, with two exceptions, of types otherwise known only from the mid-Devonic. The pelecypods are similarly mid-Devonic types, and so are the trilobites. The gastropods and cephalopods, on the other hand, are without exception upper Siluric types."

Of the Lucas fauna, Grabau says[3] " This is the highest fauna of the Monroe group, and it is throughout a Siluric fauna." An analysis of his summary of this fauna, however (page 221), gives the following information: Of corals there are two specific identifications, one species *Cylindrohelium profundum*) being new and one species (*Cladopora dichoto. na*) persisting from the Anderdon limestone; of seven species of brachiopods, four (*Prosserella lucasi, P. subtransversa, P. unilamellosa, P. planisinosa*) are found only in the upper Monroe, two (*Schuchertella interstriata* and *Spirifer modestus*) in the upper Silurian, and one (*Camarotoechia semiplicata*) in the upper Silurian and lower Devonian; of three pelecypods, two (*Panenka canadensis* and *Conocardium monroicum*) are found in the lower horizons of the upper Monroe and are of decided Devonian character, and one (*Pterinea bradti*) is a new species considered a derivative of a species of the lower Monroe (*P. lanii*); of nineteen species of gastropods listed on page 212, three (*Hormotoma subcarinata, H. tricarinata,* and *Pleurotrochus tricrinatus*) have their nearest relations in the upper Silurian of Gottland, six (*Eotomaria areyi, E. galtensis, Lophospira bispiralis, Pleurotomaria* cf *relaris,* and *Polcumita* cf *crenulata*) are Guelph, and three (*Solenospira minuta, S. extenuatum,* and *Holopea subconica* are Manlius, two species (*Pleuronotus subangularis,* and *Trochonema ovoides*) are closely related to

[1] Michigan Geol. and Biol. Surv., Pub. No. 2, Geol. ser., p. 217.
[2] Ibid. p. 221.
[3] Ibid. p. 221.

Devonian for is; and the remaining five species (*Acanthonema holopiform*, *A. holopiformis* but. *obsoleta*, *A. laxa*, and *A. Newberryi*) are confined the Lucas and Amherstburg beds.

In all, twenty-four genera are recognized in the Lucas dolomite. these, two are known only in the upper Monroe, and the remainder represented as follows: six, Ordovician to Carboniferous or later; tw Ordovician to Devonian; two, Ordovician to Silurian; one, Silurian Carboniferous; two, Silurian to Devonian; four, Silurian; and one, Devoni. Thus six genera of the Lucas unquestionably have Silurian affinitie These consist of one genus of coral (*Cylindrohelium* a new genus to whi tirabau doubtfully refers a Guelph species), and five genera of gastropo represented by species " mostly exotic in character." In view of t presence of unquestioned Devonian elements in the Lucas fauna, and the decidedly middle Devonian character of the faunas of the other upp Monroe divisions which underlie the Lucas, it seems decidedly inadvisal to place this whole formation in the Silurian on the evidence of gastropoc which are admittedly " exotic," and are of a class of organisms know to be poor guide fossils in the Palaeozoic system. It would seem mu more logical to consider the Detroit River series or upper Monroe as Devonian age, the Silurian species present being regarded as holdove from Silurian time. This is in accord with Grabau's alternative correl tion. He says[1]:

" If correlation were to be based on faunal evidence alone, a differe interpretation of the stratigraphy of Michigan would probably be adopte In that case, the lower Monroe would be correlated with the upper Ca ugan. . . . .

Faunally the upper Monroe might be considered as the indigenou lower Devonic. . . . . On this hypothesis, the Sylvania woul represent the continental condition appearing at the end of the Silu rie. . . . .''

Grabau, and Stauffer[2] as well, were greatly impressed with the erosio interval at the top of the Monroe formation, an interval sufficient to allo of the removal of the Lucas and Amherstburg dolomites from the top c the Anderdon limestone at Amherstburg and Sibley before the Onondag sediments were deposited. This interval seemed to them to . th degree of magnitude required to separate systems. It must be remem bered, however, that the formations eroded were to a considerable exten magnesian calcilutites and calcarenites, or in other words dolomites forme from the erosion of other dolomites and hence the time required both fo their deposition and for their erosion would be much less than that require for the deposition and erosion of ordinary dolomite. As there was littl time for hardening, the erosion of these dolomites would undoubtedl be rapid.

That the time interval between the lower and upper Monroe ma have been at least as long as that mentioned above is suggested by th presence, between these divisions, of the continental Sylvania sandstone Grahan states[3] that there is reason to believe that the Sylvania sandston overlaps disconformably the different members of the lower Monroe, an even rests on the Niagara formation in Indiana. This indicates a muc

[1] Ibid, p. 233.
[2] Geol. Surv., Can., Mem. 34, p. 285.
[3] Ibid, page 36.

greater erosion than that at the top of the Monroe. Again, in the character of the Sylvania sandstone good evidence of emergence and continental conditions exists, and the erosion at the top of the Monroe may so far as evidence goes, have been in part at least due to current action although the final conditions were those of emergence. Admitting, however, the difficulty in explaining the lack of sufficient sedimentary or other records which elsewhere might be expected to mark the long period required for the deposition of the upper Monroe and the erosion which removed a large part of it, it must be remembered that the difficulty is not removed but only shifted to Silurian time by considering the upper Monroe of Silurian age. The deciding element in the whole argument appears to the writer to be found in the fossils which are admitted by all to be Devonian in all essential characters.

Although Stauffer formerly preferred to place the upper Monroe in the Silurian system, he has in his later writings given up this opinion. He says:[1] " It may be that the unconformity (disconformity) at the base of the Oriskany in western New York, represents the one occurring at the base of the Detroit River series, and that the unconformity (disconformity) at the top of the Oriskany is the one so prominent at the base of the Onondaga in Ohio, Michigan, and extreme southwestern Ontario. If this be the case, these two unconformities run together at various places in Ontario near the eastern end of lake Erie, but diverge rapidly to the westward from Springvale. There is thus represented an unknown interval during which it is probable the Detroit River series was deposited; for to maintain that such a fauna as that found in the Amherstburg dolomite is Silurian is more impossible than to find a place for it among the recognized Devonian formations."

Having settled the time classification to be adopted in this memoir for the Detroit River series, the age of the Sylvania sandstone has to be considered. Grabau concludes[2] that the sand, which may have been derived from the St. Peter or Potsdam sandstone to the west, was probably in part water and in part wind-borne at first, but later, as emergence continued, it was entirely wind-borne for a time, the sand dunes being reworked by the subsequent, advancing sea. This would account for the presence of Detroit River fossils in intercalated shale beds in the upper part of the Sylvania sandstone and for the intermingling of sand and dolomite as is found to be the case in the wells of western Ontario. Such a formation is typically tangential, and different parts of it may readily be assigned to different periods of time. In Ontario, however, the last phase of sedimentation seems to be predominant, and this fact, along with the content of Detroit River fossils in lenses near its top, favour placing the Sylvania at the base of the Devonian as here represented. As suggested in Stauffer's statement quoted above,[3] the thin lenticular sandstone at the base of the Dundee at Detroit river may represent the Oriskany horizon. That the purely æolian deposition may, as stated by Grabau in his alternative correlation of the Monroe formation, be classed as Silurian (better epi-Silurian) in the sense that it occurred during continental conditions at the close of the marine Silurian deposition, is possible, although improbable, as this would pre-suppose that the Sylvania sandstone bridged the gap

[1] Bull. Geol. Soc. Am., vol. XXVII, 1913, p. 77.
[2] Ibid. page 79.
[3] Ibid. page 76.

occupied by the highest Silurian and the Helderbergian strata elsewhe
The classification adopted by the writer seems more satisfactory for t
description of Ontario occurrences.

In adopting the above classification, the terms upper Monroe a
lower Monroe are obviously no longer appropriate, since the so-call
Monroe is now seen to belong to different geological systems. Grabau
alternative names " Detroit River " and " Bass Island " will hence
used.

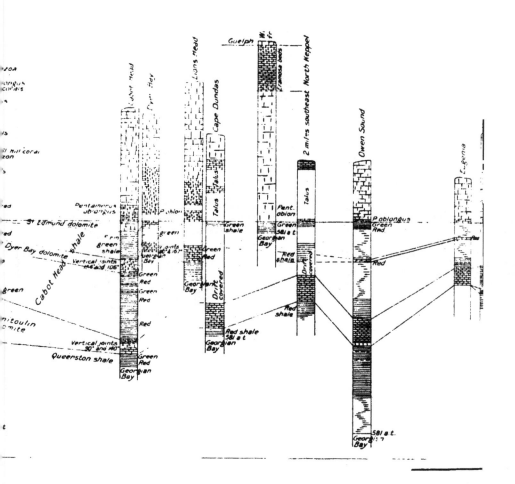

Figure 3. Diagram showing sections along the Niagara escarpment

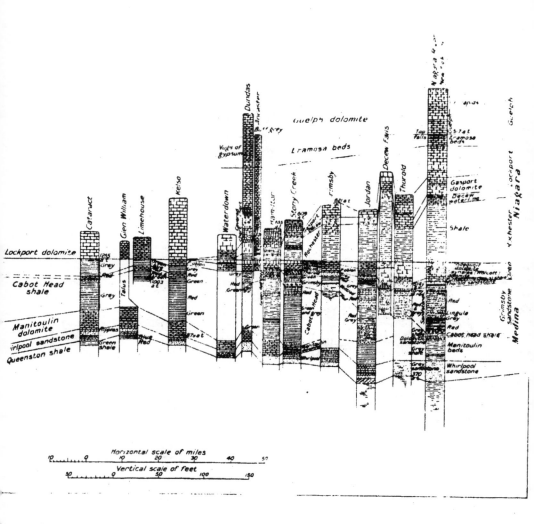

Lockport dolomite

Cabot Head
shale

Manitoulin
dolomite

Whirlpool sandstone

Queenston shale

Guelph dolomite

Eramosa beds

Horizontal scale of miles

Vertical scale of feet

Manitoulin island.

Sections prepared from surveys by M. Y. Williams, 1913-1916

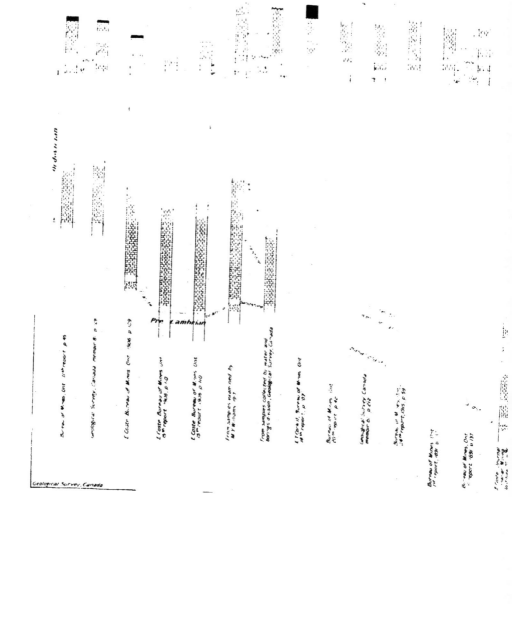

Bureau of Mines Ont 11th report p 45

Geological Survey, Canada memoir 8 p 48

E Coste Bureau of Mines Ont 1906 p 129

E Coste Bureau of Mines Ont 15th report 1906 p 10

E Coste Bureau of Mines Ont 15th report 1906 p 110

From samples examined by W V Winham 197

From samples collected by water and borings d'vison, Geological Survey, Canada

E F Caird, Bureau of Mines Ont 14th report p 07

Bureau of Mines Ont 10th report p 40

Geological Survey Canada memoir 6 p 212

Bureau of Mines Ont 14th report 1915 p 59

Pre Cambrian

Bureau of Mines Ont 11th report, 1891 p 17

Bureau of Mines, Ont report 1891 p 137

E Coste Journal Water Supply 21 1883 p 14

Geological Survey, Canada

CHAPTER V.

DESCRIPTION OF FORMATIONS.

THE MEDINA-CATARACT FORMATION.

EXPLANATION OF NAME.

The Ontario sediments of the Medina-Cataract formation vary so distinctly by lateral gradation that a hyphenated term seems best suited for the name of the formation as a whole. The term Medina, as restricted by Grabau, is applicable to the formation at Niagara Falls and may be extended to Hamilton, if minor changes are overlooked. It would thus include the sediments between the Queenston shale below and the Clinton formation above, the subdivisions in ascending order consisting of the Whirlpool sandstone, the Manitoulin beds more or less developed into dolomite (a decided dolomite at Stony Creek and Hamilton), the Cabot Head sandy shales, the Grimsby sandstone (thin at Hamilton), and the Thorold sandstone (absent at Hamilton). The term Cataract as defined by Schuchert is applicable from Dundas (and fairly applicable at Hamilton which is in the transition zone) north, and includes the Whirlpool sandstone, the Manitoulin dolomite, and the Cabot Head shale which contains in its upper half on Bruce peninsula and Manitoulin island the Dyer Bay and St. Edmund dolomites. The littoral deposition of sand forming the Whirlpool sandstone terminated southwest of Collingwood and the Grimsby and Thorold sandstones grade northward into mudstones and sandy shales of the Cabot Head shale division, probably because of the lack of sandy material due to the distance from its source. As may be seen from Figure 3, the Cataract rests successively to the north upon the Queenston and marine Richmond sediments and is overlain successively by the Clinton and the Lockport dolomites.

GENERAL DESCRIPTION.

The term " Medina," as already stated, is here used in the restricted sense defined by Grabau,[1] that is to include the "upper Medina" of the older reports or the sandstones and shales between the Queenston shale below and the Clinton formation above. Believing that the Queenston shale is separated from the " upper Medina " by a systemic boundary, Grabau found it impossible to use the term Medina to include the " upper Medina " and the Queenston shales as had formerly been done, and he concluded that the term was first referred to the section near Medina, New York, where the upper division is the prominent one. The term Cataract includes the so-called " Ontario Clinton " and the Whirlpool sandstone, or the sandstones, dolomite, and shale between the Queenston, or in the north the Richmond, below, and the Clinton, or farther north the Lockport, above. From Hamilton north " Cataract " is used exactly as defined by Schuchert. the author of the term. The present usage,

[1] Science, new ser., vol. XXIX. 1909. p. 356.

not the conception of the facts, however, differs from that of ?
in the Niagara peninsula, where Schuchert restricts the ter?
to the Grimsby and Thorold sandstones (described below), using
Cataract for the lower members of Grabau's Medina. There i
accord, however, in the belief that the Grimsby and Thorold s:
are only facies of the top of the Cabot Head shale occurring farth
although they represent another faunal invasion. Briefly, the
contention is that the term " Medina " should be used for as mu
for ration for ? erly included as is possible, and as restricted by
is applicable throughout the Niagara peninsula, and that the term
is entirely necessary and is properly applied to the sections from I
north where the phases of sedi ? entation exclusive of the Whirlpo
stone are so different from that of the Medina that the sedime
formerly classed as " Clinton ". In referring to the whole of sout
t): , the hyphenated ter ? , Medina-Cataract, seems ? ost sati
The na ? es of sub-divisions, like Cabot Head shale, may, how
used wherever the distinguishing cl? ructers are present.

As already stated the Cataract for ? ation above the Whirlpo
stone was for ? erly thought to be Clinton. This was because
following considerations: the thinning out of the Clinton li ? esto
the overlying Rochester shale fro ? Niagara river westward; the
? ent of the Manitoulin dolomite in the sa ? e region; the disap
of the Thorold sandstone or Grey Band of New York, in the vi
Hamilton; and the wrong interpretation of the Whirlpool sand
the "Grey Band" fro ? Ha ? ilton north. These mistakes are mo
evident in the "Palæontology of New York" vol. II, 1852, and
evident in the "Geology of Canada", 1863, and Logan, to so ? e ext
the difficulties involved. On page 322¹ he says "The bluish-blac
which in the state of New York afford a well-? arked division betv
Clinton and Niagara for ? ations, are available for this purpose,
a short distance in Canada...... We, therefore, propose to in
the Niagara series, the two bands of li ? estone which unde
shales, and which, in New York, constitute the upper part of the
for ? ati ? n." The Cataract fossils, particularly those collected at H
and Dundas (e.g. *Cælospira planoconvexa*, etc.) were figured by Ha
with the true Clinton fossils, and subsequently palæontology was
back up the error in stratigraphic determination. Confusion reigne
transition zone between Hamilton and Niagara river where b
Medina and Cataract characters are fairly well-? arked, but fro ? H
north the dolomite and shale of Medina age were unhesitatingly
as "Clinton". The "Clinton" of Ohio, in part thought to corre:
age with the Cataract, for ? ed another ano ? aly and it was only :
tangle was straightened out by Schuchert for Ontario, that har ?
finally attained. Such differences as may exist in recent publicat
rather ? atters of ter ? inology than of interpretation of facts. The
study of the northern sections by the author in 1915-16 has adde
tional infor ? ation relating to the top of the Cataract section :
corrected interpretations of the author which were embodied in Sch
account of the Cabot Head region.

¹Palæontology of New York, vol. II.

Following is a description of the members of the Medina-Cataract formation, taken in ascending order.

WHIRLPOOL SANDSTONE[1] MEMBER.

Extent and Characteristics.

From Niagara river to the vicinity of Duntroon, about 8 miles south of Collingwood, the Whirlpool sandstone is the basal member of the Medina-Cataract formation, and so the oldest representative of the Silurian system. The typical sandstone consists of fine quartz grains, is light grey in colour, and is in part crossbedded. At some localities, notably in the vicinity of Credit Forks, chocolate-brown beds occur interbedded with the grey. From a thickness of 25 feet at Niagara river and 30 feet reported from a well boring at Thorold, the Whirlpool sandstone thins to the west and north, being 4 feet thick at Glen Huron and 6 feet thick, 3 miles west of Duntroon. At Mitchell's mills, 2 miles west of Kolapore, 6·5 feet of green shale with sandy beds at the base, may represent an off-shore occurrence of the Whirlpool sandstone. This is the extreme north-westerly occurrence known to the writer.

West of the Niagara escarpment, the information relating to the extent of the Whirlpool sandstone is derived from records obtained from well borings. In spite of the devious methods by which many of these records have been kept, a fair idea of the geographical distribution of the sandstone may be obtained from them. Thus we learn that no sandstone occurs at the top of the Queenston west of the city of Guelph, neither is any present at Kitchener, London, Petrolia, nor in Kent county. At Beachville, calcareous sandstone is reported, which may represent the Whirlpool sandstone, in part reworked at a later time. In Colchester township, Essex county, 62 feet of sandy limestone occurs at the Manitoulin dolomite horizon and the sand content may represent the Whirlpool deposition. In Norfolk county, the Whirlpool sandstone is reported to be 5 feet thick at Port Burwell, 23 feet at Vienna, 27 feet at St. Williams, 10 feet at Vittoria, 18 feet at Port Ryerse, 12 feet at Port Dover, and 3 to 5 feet in the vicinity of the town of Simcoe. Thicknesses are reported elsewhere as follows: in Haldimand county, between 10 and 30 feet; in Welland, 19 feet at Port Colborne and 30 feet at Thorold; and near Brantford between 10 and 20 feet. From the above evidence (omitting the uncertain evidence from Essex county) the western underground extension of the Whirlpool sandstone may be limited within our knowledge by a line drawn from a point near Duncan, 13 miles southwest of Collingwood, southeasterly past Heatherton and thence southerly between Guelph and Kitchener, and southwesterly to lake Erie near Port Stanley.

The Whirlpool sandstone everywhere rests on the red Queenston shale, which is almost universally mud-cracked at the top, the sandstone projecting down into, and filling the cracks. In general, the lower half of the sandstone is thick-bedded to massive, the upper half is thin-bedded, and, where overlain by shale, grades into the overlying beds, through a

Jour. Geol., vol XVII, No. 3, p. 238.
North and west, from the vicinity of Dundas, the Whirlpool sandstone was described by Sir Wm. Logan (Logan 1863, p. 313) under the name of "Grey band." The true "Grey band" of former writers, now known as the Thorold quartzite, does not occur to the west and north of Hamilton and as a consequence, Logan confused the more northern occurrences of the Whirlpool sandstone with it

transition zone of alternating sandstone and shale. Crossbedding observed in some part of almost every section, but is confined to a limite thickness. Shale lenses occur here and there in the crossbedding. Wit the overlying Manitoulin dolomite the contact is sharp, but the lowe dolomite beds are arenaceous. At Glen William, ripple-marks occur i the upper bed of sandstone, and at Credit Forks oval masses of sandston are enclosed in shale coverings, which are in many cases full of worm burrows. The upper beds of the sandstone here are undulatory, an "pillow" structure occurs locally, at the very top. One specimen o "pillow", about 2 feet long, is shaped like a sheaf of wheat with shal containing casts of worm burrows or fucoids wrapped around its sides an partly over one end. Ripple-marks occur on some of the "pillows". Th base of the overlying dolomite is full of vugs of pink celestite.

Under the microscope, typical Whirlpool sandstone taken from on of the "pillows," is seen to be made up of angular to subangular quart grains of 0.1 mm. average diameter. No secondary quartz was observed but there is recrystallization indicated by wavy extinction, along th margins of the sand grains, which are closely compressed, in interlockin arrangement.

Fossils are of rare occurrence in the Whirlpool sandstone and ar found for the most part in the upper beds. The following have bee obtained by the writer and his assistants. *Pleurotomaria* sp. (fragmentary) *Hormotoma subulata* Hall? ?*Cornulites*, and casts of worm burrows o castings and fucoids (see Table and Plates II, III, and VIII).

Origin.

The origin of the Whirlpool sandstone is a subject of some interest A. W. G. Wilson[1] has described the Whirlpool sandstone ("grey band" of Ontario) as of probable sand dune origin, and cites the mud-cracked character of the upper Queenston shales, and the crossbedding of the Whirlpool sandstone itself, in support of his contention.

Grabau[2] derives his "Medina red sands" and "Thorold quartzite" from the erosion of the folded Juniata beds and Bald Eagle quartzites o the Appalachian region. He says of the Whirlpool sandstone.[3] "A. W. G Wilson has interpreted this lens of sandstone as the remnant of an old dune area, reworked by the advancing sea, and this interpretation is favoured by the character of the grains composing this rock as well as by its distribution Some of the bedding surfaces of this quartzite show wave-marks closely resembling those of modern shallow beaches, a feature repeated in a number of higher beds farther east. The Whirlpool quartzite is not known a Rochester . . . . The Whirlpool quartzite is thus seen to be a loca formation apparently unconnected with any direct eastern source."

That the shales and sandstones of the Queenston and Medina bed were derived in general from the Appalachian region, there seems littl reason to doubt, and it appears to the writer that though the "Whirlpoo quartzite" may be "apparently unconnected with any direct eastern source" its deposition may be readily explained as the result of geological processe at work after Queenston time.

[1] Can. Rec. of Sc., vol. IX, No. 2, 1903, pp. 120-122.
[2] Bull. Geol. Soc. Am., vol. XXIV, 1913, pp. 414-415
[3] Id. at page 461.

Grabau has hypothecated the erosion of Bald Eagle beds to supply the material of the Thorold quartzite. No good reason is apparent why these beds may not have been eroded to some extent at least, before the Thorold deposition. As might be expected, the Queenston shales are more arenaceous in the east, nearer their source, and it is more than probable that sand bars of river origin, flood-plain deposits of gravel and sand, and possibly piedmont deposits of coarse materials characterized their landward extension during their period of deposition.[1] The nearby old land might be expected to furnish such deposits. With the return of shallow water conditions indicated by mud-cracks in the Queenston shales of western New York and Ontario, rejuvenation of the erosion processes would follow, and though shale would be carried farther seaward, a littoral zone would receive the sandy deposits. Wind may have played a large part in the distribution of the sand over the mud-cracked Queenston surface or gentle currents may have been the distributing factor. Again, the cleanly sorted, even-grained nature of the Whirlpool sandstone, although due no doubt in great measure to its long transportation by stream action and to subsequent wave action in the littoral zone, may be dependent to some extent upon the effective sorting done by the wind. The fact that from Rochester eastward the Whirlpool sandstone has not been recognized, may be explained on the ground that sorting has been less effective and consequently the red sandstone of contemporaneous age with the Whirlpool cannot be distinguished from similar beds above and below. It is difficult indeed to determine the boundary between the Queenston and Medina sandstones from Rochester eastward.

Thus the origin of the material of the Whirlpool sandstone very probably dates back to the emergent period marking the interval between Ordovician and Silurian time. That the Whirlpool sandstone has received its present form during the advance of the sea at the commencement of Silurian time is shown by the following characters: (1) the formation has a comparatively uniform thickness; (2) crossbedding is limited to a part of the beds only and is more regular than is that characteristic of wind deposition; (3) the upper beds are generally thin, and are interbedded with shale or contain pillow structure; (4) the upper beds are sparingly fossiliferous, the fauna being characteristic of the marine littoral zone.

Shale occurs in the crossbedded sandstone at Credit Forks, but this may be the case even in aeolian deposits as shown by Grabau.[2]

As will be shown later, subaerial distribution of the Whirlpool sand over the semi-emergent Queenston surfaces may readily have resulted in alternating sand and shale beds on the invasion of the Silurian sea. Easterly trade winds at this time, as suggested by Grabau,[3] would appear to fulfil the required conditions. Whether such winds existed or not the emergence of the land would readily account for seaward extension of the coarser clastic deposits as described by Wilson.[4]

[1] Bull. Geol. Soc. Am., vol. XXIV, No. 3, 1913, p 464.
[2] Principles of stratigraphy, p 564.
[3] Grabau, A. W., "Palæozoic delta deposits," pp 440 et seq.
[4] Wilson, A. W. G., Can. Rec. Sc., vol. IX, No. 2, 1905, p. 113

Fauna of Whirlpool Sandstone.

Gastropoda:
Pleurotomaria sp. Niagara River;
Hormotoma subulata (Conrad)?? Glen William.
Vermes (burrows and trails).
?Cornulites.

MANITOULIN[1] MEMBER.

General Description.

At its typical occurrences, on Manitoulin island, the Manit
member which rests on Richmond shale and is overlain by Cabot
shale is a resistant, thin to thick-bedded, grey or buff-weathering dolo
which when fresh, is hard and of light, blue-grey colour. Near the
the dolo ite is thin-bedded and argillaceous; midway up beds 10 f
ore in thickness occur; and thin beds are again present near the top.
escarp ent of Manitoulin dolo ite, about 1 ile east of the hea
Manitowaning bay, easures 50 feet in height, so that allowing for g
tion and other erosion, the total thickness here ust have been at
60 feet.

In the vicinity of Manitowaning, local coral and bryozoan
occur in the upper 20 feet of the dolo ite. The reefs are several
in dia eter, and cause local thickening of the dolo ite and
structure.

At Cabot Head lighthouse, the Manitoulin dolomite is about 10
thick, but beds occur to the westward which give indication tha
total thickness is 25 to 30 feet. At Owen Sound, 32 feet of this dolo
is exposed in a quarry, the rock being thin-bedded, and of grey c
with blue shale partings. Fro 23 feet at Credit Forks, the thickne
the dolo ite decreases to the southeast to 8 feet at Stoney Creek, be
which the dolo ite gives place to calcareous shales. At Niagara g
a 3-foot bed of calcareous sandstone rests on 25 feet of shales.
sandstone appears to represent the southeastern extension of the u
ost of the Manitoulin beds.

Well records give the Manitoulin dolo ite as 15 to 20 feet thic
Brantford; 42 feet thick in Pilkington, and 29 feet in Peel town
Wellington county; and 50 to 60 feet thick between Allenford and Wia
At Woodstock 35 feet of calcareous sandstone is reported and in the we
part of Kent county the Manitoulin dolo ite is 20 to 50 feet thick.
a record of a well in Colchester South township, Essex county, 62 fe
"grey blue and white sandy li estone" is reported fro the Manit
horizon. It is conse uently fairly certain that excepting the alca
shales and sandstones at the Manitoulin horizon in the eastern and sout
part of the Niagara peninsula, the Manitoulin dolo ite underlie
whole of the Silurian area of southwestern Ontario. Where such
are not reported it is by no eans certain that they are absent, as
may easily have been overlooked in keeping the well records.

[1]Proposed by author, Ottawa Nat., vol. XXVII, 1913, p. 37. Formerly proposed by A. C. Lane, Geol
Mich., vol. 1, 1873, as a group n e to include Niagara and Guelph This usage is not generally recognized.

In the south, the Manitoulin beds rest on the Whirlpool sandstone. Even where the basal beds are shale, the contact is well-defined, but transitional sandstone and arenaceous shale alternate above the top of the well-defined Whirlpool sandstone. Where dolomite rests on the sandstone, the contact is generally sharp, but the lower dolomite beds are sandy or arenaceous. The dolomite rests with sharp, even contact on red Queenston shale from Collingwood west and north to the end of Bruce peninsula, the Queenston last being seen at Cabot head. On Manitoulin island the contact is with green Richmond shale. This contact is sharp, but the basal Manitoulin dolomite is somewhat argillaceous. The two faunas have practically nothing in common.

Microscopic Characters.

A specimen of fresh Manitoulin dolomite from a quarry at Owen Sound, proves under the microscope to be composed of interlocking grains of dolomite of 0·12 mm. average diameter.

Two specimens of the dolomitic sandstone of Niagara river were examined microscopically. One proved to be made up of rounded quartz grains of 0·13 mm. average diameter, with about 50 to 55 per cent dolomite. The other is composed of rounded to subangular quartz grains of 0·11 mm. average diameter, and 5 to 15 per cent of dolomite. One, small, worn fragment of plagioclase was observed in each slide, the size corresponding to that of the average quartz grains.

The arrangement of the crystals of dolomite indicates that they have crystallized around and have filled in pore spaces between the quartz grains.

Under the high power, the edges of the quartz grains are found in many cases to be gently etched, indicating some solution by the highly charged dolomitic waters which surrounded them.

Fossils.

As will be seen from the table (page 30 and Plates III, V, VI, and VII) the Manitoulin dolomite is for the most part decidedly fossiliferous.

Origin.

Of the origin of the Manitoulin dolomite, Grahan says[1]: "The Keppel (previously named Manitoulin by the author) dolomite is an earlier deposit before the sands and muds from the Appalachians reached this point, or during the earlier period of the erosion of the folded red beds of the Ordovicic." Schuchert[2] says "the Cataract province spreads westward through the Saint Lawrence embayment", and concludes "that the normal marine junction of the Cataract and Brassfield seas is prevented by the Medina delta".

[1]"Palæozoic della deposits," p. 460 cf. 469.
[2]"Medina and Cataract formations of the Siluric of New York and Ontario"; Bull. Geol. Soc. Am., vol XXV, p. 294.

Fauna of Manitoulin Dolomite.

Genera and species.	Niagara River.	Credit Sy.	Stoney Creek.	Hamilton.	Dundas.	Glen Williams.	Cataract and Credit Forks.	Lavender Hills.	Banks.
Plantæ									
Buthotrephis gracilis crassa Hall									
Hydrozoa									
" Inthrodictyon vesiculosum Nicholson and Mur.									
Anthozoa									
Streptelasma f huskinsoni Foerste									
Enterolasma cf geometricum (Foerste)									
Zaphrentis bilateralis (Hall)									
Palæophyllum umbellifererens Chadwick									
Palæophyllum (Cyathophylloides?) williamsi sp. m.									
Amplexus okunardi (Edwards and Haime)									
Chonophyllum ? ali Billings									
rularia gracilis (Billings)									
Ceratopora tenella (Renninger)									
S. fragopora farmis Billings									
Palæofavosites asper D'Orbigny									
Lyoites catenularia niercopora Whitfield									
Cyathoidea									
Brorkocystis tecumseth (Billings)									
Crinoidea									
Eucheiocrinus ontario Springer n. sp									
Bryozoa									
Homotrypa? ex labs (Foerste)									
Hallopora magnopora (Foerste)									
Helopora galis Hall									
Pachydictya turgida Foerste									
Brachiopoda									
bala cuneata Conrad									
Strophomedia striata Hall									
Leptæna rhomboidalis (Wilckens)									
Plectambonites transversalis Wahlenb.									

Hebertella fausta (Foerste)

Platystrophia biforata (Schlotheim)

Dalmanella eugeniensis Williams n. sp.

Dalmanella eugeniensis paleoelecantula Williams n. var.

Rhipidomella hybrida (Sowerby)

Camarotoechia (Stegerhynchus) neglecta (Hall)

Rhynchotreta cuneata americana (Hall)

Rhynchotreta williamsi Foerste n. sp.

Rhynchotreta raboteensis Williams n. sp.

Rhynchonella? janea Billings

Rhynchonella? bidens Hall

Atrypa parksi Williams n. sp.

Whitfieldella cataractensis Williams n. sp.

Cyclospira planoconvexa (Hall)

Gastropoda

Bucania trilobata (Conrad)

Bucania cf exiguus Foerste

Lophospira pulchra Williams n. sp.

Strophostylus cancellatus (Hall)

Hormotoma subulata (Conrad)

Cephalopoda

Orthoceras ignotum Foerste

Orthoceras hamaverensis Foerste

Orthoceras multiseptum Hall

Vermes

Cornulites distans Hall

Cornulites incurvus (Shumard)

Trilobita

Calymene clintoni (Vanuxem)

---

[1] Reported by Parks, Twelfth Inter. Geol. Cong. Guide Book No. 7, 1913. p. 10.

[2] Identified by Bassler.

Although described by the two writers from widely divergent ⟨
points, one idea is back of both statements, that is, an encroaching
from the northward, tending to deposit calcareous material, was inv⟩
by detrital material from the Appalachian province. The greater t⟨
ness of the Manitoulin dolomite in the north and the sandy shaly ph⟩
of this member in the south, indicate quite clearly the encroachin⟨
fairly deep, marine conditions over the littoral zone occupied by⟩
Whirlpool sandstone. That the supply of detrital material became gr⟨
in post-Manitoulin time than the deposition of calcareous deposits is ⟨
by the shaly or sandy characters of the subsequent members of the Cata⟩
formation.

CABOT HEAD SHALE MEMBER.

General Description.

The Cabot Head shales, as already explained, were formerly inclu⟨
with the Manitoulin dolomite in the Clinton formation of Ont⟩
Schuchert[1], in 1912, included these shales in the Cataract formation⟨
in 1913 the author[2] proposed for them the name "Kagawong meno⟨
or "Kagawong shales." This name was preoccupied by Foerste⟨
had applied it to an upper Richmond member. "Cabots Head"⟨
proposed by Graban[3] in 1913 for the red and grey Cataract shales⟨
included dolomites of the mainland of Ontario. In its corrected f⟨
"Cabot Head" (following the recognized form of the place name on maps⟨
charts), the author finds it applicable for the Cataract shales of southw⟨
ern Ontario and the Georgian Bay region.

Detailed Description and Sections.

The Cabot Head shales vary much in character in different part⟨
the same section, as well as at different localities. They include ⟨
and red shales and fine-grained, shaly sandstones, ferruginous, sa⟨
dolomite, and grey, shaly dolomite. At Niagara river, where some sa⟨
dolomitic beds and grey shales below them are classed as a southern fa⟨
of the Manitoulin dolomite, only 3 or 4 feet of grey shale are classified⟨
Cabot Head. Here, the Grimsby sandstones are thick and character⟨
tically developed. As they become shaly to the north they add to⟨
thickness of the Cabot Head shales of which they are a southern fa⟨
The general colour of the shales from Niagara river to Wiarton is g⟨
but red, sandy beds occur in the upper half including a red, sandy, d⟨
mitic bed well exposed at Kelso, Limehouse, and Cataract. The sh⟨
of Bruce peninsula and Manitoulin island contain red beds in their l⟨
half and beds of dolomite interbedded with grey shale in their u⟨
half.

At no place do the Cabot Head shales show more variations t⟨
at the type locality of Cabot head. The nearly flat-lying Manito⟨
dolomite at Wingfield basin, upon which Cabot Head lighthouse sta⟨
dips gently to the southwest beneath the escarpments of three promi⟨

[1] Bull. Geol. Soc. Am., vol. XXIV, 1913, p. 107
[2] Ottawa Naturalist, vol. XXVII, pp. 37-38
[3] Bull. Geol. Soc. Am., vol. XXIV, No. 3, Sept., 1912, p. 450

bluffs, known locally as Bowlder, Middle, and West bluffs. Talus obscures much of the section, but as worked out it consists in ascending order of 70 feet of shale, red near the base and containing firm beds bearing *Halopora fragilis*, and firm and green near the top: 4 feet of soft green shale; 15 feet of thin, even-bedded dolomite (Dyer Bay); 32 feet of firm green shale; 8 feet of rather massive, creamy weathering dolomite (St. Edmund); and 4 feet of soft, green shale. Lockport dolomite succeeds, the lowest 8 feet being thin-bedded, green in colour, and in part mud-cracked.

The above section was described by Logan[1] and copied by Grabau[2]. Logan's description follows: "At Cabot's Head, the very summit of the Medina formation "(Queenston)" is seen at the water's edge; and there rest upon it about twenty-six feet of a dolomite . . . . . . On the dolomite, repose 103 feet of red, marly sand-stones, partially striped and spotted with green, and interstratified with beds of red and green argillaceous shale; none of which exceed six or eight inches in thickness. The green argillaceous beds appear to be quite free from calcareous matter, and the stone is carved by the Indians into tobacco pipes. These red and green strata are succeeded by about fifty-five feet of green calcareo-argillaceous shales and thin-bedded limestones, and terminated by the massive limestones of the Niagara series."

Three miles west of Cabot head, a fine exposure, locally known as the "Clay cliffs,"[3] occurs on the shore (Plate XXXIA). The Manitoulin dolomite is here below the water, the section being, in ascending order from the water til, as follows: lower 22 feet obscured by talus, probably red shale; 5 feet of soft, red shale; 17 feet of soft, red shale, interbedded near the base with thin calcareous shale, and containing also firm shale beds to about one-fifth the total amount; 3 feet of green clay shale; 15 feet of massive to thin-bedded dolomite (Dyer Bay); 36 feet of thin-bedded, firm, green shales; 8 feet of buff-weathering dolomite (St. Edmund) containing *Favosites cristatus*; 14 feet of covered interval, succeeded by 4 feet of dolomite belonging to the Lockport formation and bearing *Pentamerus oblongus*. From a comparison with the Cabot Head section it is evident that the upper 8 to 10 feet of the covered interval includes thin dolomites belonging to the base of the Lockport formation and that only the lower few feet are shale beds.

The upper dolomites (" limestones ") were clearly included by Grabau in the Cabot Head shales, and so there appears no good reason for redefining the term. Ferruginous limestone is found near the top of the Cabot Head shales at Credit Forks and Limehouse, and thin-bedded dolomites occur interbedded with shale in the upper part of the Cabot Head beds

[1] Geology of Canada, 1863, pp. 312-320.
[2] Ibid, page 400.
[3] On seeing the section at the Clay cliffs, for the first time, the author confused the 15 feet of dolomite in it with the Manitoulin dolomite at Cabot head, and as a covered interval at the base of Centre and West bluffs suggested a structural break, he stated in the Summary Report of the Survey for 1913 (p. 183) as follows: "Cabot Head was proposed by Grabau for the shale under discussion and it is here accepted. The writer, however, wishes to draw attention to the fact that the measurements and description of the section at Cabot head, which were taken from Logan, are seriously in error, due to the failure on the part of former workers (and the writer also upon a previous occasion) to recognize a structural disturbance, which repeats much of the section." For the same reason the thickness of these shales is given in the Summary Report for 1912 (p. 279) as 50 feet. Likewise the plate published by Schuchert, (a) which was taken from a photograph by the author, is wrongly described in the title. In reality it is the 15-foot bed of dolomite (Dyer Bay) high up in the Cabot Head shales. The true Cataract (Manitoulin) dolomite is here submerged. The section given by Logan, though differing slightly in measurement, is substantially correct, as may be seen by a comparison with the section by the author given above.

a) Bull. Geol. Soc. Am., vol. XXV, 1914, pl. 14, fig 2, opposite p. 287.

on Manitoulin island. The dolomite beds at Cabot head appear worth of some designation, particularly as they carry, as will be described late characteristic fossils. It is proposed to designate the lower, the Dy Bay dolomite lentille, and the upper, the St. Edmund dolomite lentil The lower dolomite contains numerous characteristic fossils at Dyer B and St. Edmund is the township in which the clay cliffs are situated an there the upper dolomite is best exposed. These dolomites are describe in detail below.

The Cabot Head shales on Fitzwilliam and Manitoulin islands a in general soft, red, clay shales in their lower beds, but contain soft, gre shales, firm, green shales, and dolomites (Dyer Bay) near the top. The at the northeastern extremity of Fitzwilliam island, 15 feet of soft, re clay shale is overlain by 6½ feet of soft, green shale and thin, firm, gre shale interbedded. West of Tamarack cove, in a partial section, sof green shale is overlain by 15 feet of green-grey, fine-grained dolomi (Dyer Bay) which is succeeded by 22 feet of firm, pea-green shales. covered interval obscures the beds above.

In the partial sections along the east side of lake Manitou son firm, green shale occurs above 10 to 15 feet of exposed, red shale, whic where exposed at another place, overlies firm, green shale. On the roc between West bay and Kagawong, the section from the Manitoulin dol tutite up is as follows: 40 feet of soft red shale; 1½ feet of hard, gree shale, calcareous at the base; 10 feet of soft, red shale; 5 feet of soft, gree shale; succeeded by fine-grained Dyer Bay dolomite. Red shale w also observed one-half mile west of Ice ake along the road to Gore B. (27 feet of shale exposed); near the site of the former Indian village Campbell bay; and at water level at the end of the point between Campb and Helen bays. Red clay is also reported along the east side of la Wolsey.

Southward from Cabot head, both the Dyer Bay and St. Edmu dolomites extend to Dyer bay; the Dyer Bay is seen at Gun point a both are missing on Colpoy bay. Front Owen Sound east and sou as already stated, the lower Cabot Head shales are green or grey, t upper red, containing ferruginous limestones. Thin, calcareous be occur throughout the section, and arenaceous (sandy) beds and th sandstones are common in the southern sections. The Cabot He shales are 110 feet thick at Owen Sound, 60 feet at Cataract, 42 feet Hamilton, 65 feet at Stoney Creek, 63 feet at Jordan, and 5 feet at Niaga river. At the last-named locality, the beds consists of grey shale a thin sandstone, the grey and red sandstone above being included in t Grimsby sandstone which appears to be clearly the sandstone phase the upper Cabot Head shales occurring to the west and north.

Depending upon the evidence obtained from well records, there from 15 to 65 feet of Cabot Head shales at Port Burwell, 65 feet at Bea ville, 30 at Glen Allen (24 miles northwest from Guelph), 75 to 100 f east of lake St. Clair, and 100 feet in Colchester township, Essex coun Although some well logs do not clearly show the presence of Cabot He shale, enough has been learned to make it seem probable that the Cab Head shales extend (underground) to the southwestern lim of Ontari

For the fossils of the Cabot Head shales see th Plates IV, V, and VII.

Fauna of Cabot Head Shale Exclusive of Dyer Bay and St. Edmund Dolomites.

	Cabot Head	Jones Lion Creek Head	Keb Lion head	In cent

Plant
  Licrophycus verticillatus Hall
Sponge
  Receptaculites ... undosus Vall ...
Corals
  Pterodasma hexameracum ...
  Streptelasma (Checkittson) Foerste
  Zaphrentis trilineata Hall
  Palæofavosites ... et Le Orleans
Crinoids
  Delopera fragili ...
  Rhonopora verru ... Hall
Bryozoa
  Stropomella streata Hall
  Leptæna rhomboidalis Wilcken
  Hebertoniæta ... ... Whitfield
  Ortho thatellæ Foerste
  Hebertella mnista ... Foerste
  Billystrophia laforata Schloghenn
  Plectonella eugenes Walter ... ...
  Rhipidomella hybrida Foerste
  Camarotœchia ... superba ... neglecta Hall
  Rhynchonella plicata Hall
  Rhynchonella pisca Billings
  type ... ... ksi Williamsen ...
  Whitfieldella nitrametuens W... rosso sp
  ret ... per plano convexa Hall
Pelecypods
  Pterina ... rundata Hall
  Pterinæa lineæ Hall
Gastropods
  Cornulites durina Hall
  Conulites incurvus Hunter
  Nettl...
  Ctenodonta lebeodens Savage

Dyer Bay Dolomite Lentille.

Description. The Dyer Bay dolomite is typically an impure argillaceous, rather thin-bedded, dolomite lentille in the lower part of the upper half of the Cabot Head shales. Like the Manitoulin dolomite, this dolomitic deposit is a northern representative of shaly deposits farther south. At Dyer bay, the type locality, the section above water consists of 10 feet of thin-bedded, dolomite with blue (green-weathering) shale partings. Fossils are numerous and ripple-marks are common. The rock is divided by two sets of vertical joints, with directions about 342 degrees (north 18 degrees west) and 85 degrees (north 85 degrees east). About 17 feet of similar dolomite occurs in the section at the northern extremity of Lion Head cape and partial sections occur on Ba... ow and H... bays.

At Cabot head and the "Clay bank" 3 miles to the west, the complete ... is exposed. This is 15 feet thick, and consists of thin, even-bedded (appearing massive where little weathered), greenish grey, argillaceous dolomite. Characteristic ostracoda occur at the top of the section (see table). About 10 feet of this dolomite connects Centre and West bluffs, at Cabot head, as a ledge or shelf. The bedding is 1 to 3 inches thick and the formation is cut by two sets of vertical joints, with direction 149 degrees (south 31 degrees east) and 281 degrees (west 11 degrees north).

Ten feet of thin-bedded dolomite occurs in the section at the northeastern extremity of Fitzwilliam island, and in these beds *Virgiana mayrillensis* Savage occurs plentifully, along with *Schuchertella curvistriata* Savage, etc. (see table). From Fitzwilliam north and west, *V. mayrillensis* is common in the Dyer Bay beds, but generally at one definite horizon. At Tamarack cove the dolomite is 15 feet thick, rests upon green shale, and is overlain by firm, green-weathering shale. *V. mayrillensis* and the characteristic ostracoda occur here. On the southeast side of lake Manitou, about 12 feet of Dyer Bay beds occur containing *V. mayrillensis*. The thickest known section of Dyer Bay beds is exposed along the road between West Bay Indian village and Kagawong, just north of a lakelet, on lot 17, concessions VIII and IX, Billings township, Manitoulin island. Here, at least 25 feet of dolomites with some interbedded shale are included in the Dyer Bay beds. A few inches of grey, semicrystalline dolomites occur near the middle, *V. mayrillensis* occurring just below and the Dyer Bay ostracoda 6 or 7 feet higher up. West of Ice Lake on the road to Gore Bay, 5 feet of calcareous shale occurs at the top of 20 feet of red shale containing some firm green beds. In the upper 5 feet of shale, the Dyer Bay ostracoda occur at the top and *V. mayrillensis* occurs near the base.

The Dyer Bay beds are well developed and extend as ledges considerable distances from the base of the Lockport dolomite between the west side of lake Manitou and the western end of Manitoulin island, and also on Cockburn island. Typical outcrops occur on the north end of the island in Mindemoya lake, on the west side of Kagawong lake, near West bay and Ice Lake as mentioned, at the west side of Campbell bay and at Pitman point, Cockburn island. These outcrops almost invariably contain *V. mayrillensis* which was thought throughout the field work to be a small form of *P. oblongus* marking the base of the Lockport formation as the typical *P. oblongus* fills the basal beds of the Lockport dolomite at Owen Sound, Dyer bay, etc. As a consequence the Dyer Bay dolomite was mapped with the Lockport. This error has been corrected so far as could be done in the office. The true Cataract age of the Dyer Bay beds is indicated by Bassler's determination of the ostracoda and by the identification of Alexandrian fossils in the beds on Fitzwilliam island.

Southward from Colpoy bay, the typical Dyer Bay beds are absent, but 2 or 3 feet of ferruginous limestone occurs a little above the centre of the Cabot Head shale as far south as Kelso. This has been termed the " iron band " and contains common Cataract fossils, among others being *Rhynchonella? janea* which also occurs in the Dyer Bay dolomite. It seems probable that this is the southern representative of the Dyer Bay dolomite of the north (for the chemical composition of the iron band at Limehouse, see table page 111).

The recognized fauna of the Dyer Bay dolomite, as shown in the accompanying table, includes one species of plants, five of corals, nine of brachiopods, one of pelecypods, four of ostracods, and one of trilobites. Of these, two brachiopods and three ostracods are new species found only in the Dyer Bay dolomite (see Plate VII).

Fauna of the Dyer Bay Dolomite.

Genera and species.	Barrow Bay.	Dyer bay.	Cabot head.	Fitzwilliam island.	Tamarack cove.	South bay.	West bay.	Ice Lake.	Helen bay.	Cockburn island.
Plants.										
Buthotrephis gracilis Hall			x				x		x	
Anthozoa.										
Enterolasma cf geometricum (Foerste)	x									
Streptelasma cf hoskinsoni Foerste....		x	x							
Syringopora retiformis Billings			x							
Favosites cristatus Edwards and Haime	x	x	x							
Favosites obliquus Rominger......		x	x				x			
Brachiopoda.										
Strophonella striata (Hall),........		x				x				
Schuchertella curvistriata Savage.				x						
Platystrophia biforata (Schlotheim)....				x						
Virgiana mayvillensis Savage				x	x	x	x	x	x	x
Rhynchotreta thebesensis multistriata Savage				x						
Rhynchotreta cabotensis Williams n. sp ..			x							
Rhynchonella? janea Billings.	x		x				x			
Rhynchonella? bidens Hall. ..	x									
Homoeospira cf immatura Savage .				x						
Pelecypoda.										
Pterinea cf undata (Hall)		x						?		
Ostracoda.										
Leperditia cabotensis Ulrich and Bassler		x			x					
Chilobolbina punctata Ulrich and Bassler .		x								
Zygobolbina williamsi Ulrich and Bassler..		x			x					
Trilobita.										
Calymene cf clintoni (Vanuxem)	x									

St. Edmund Dolomite Lentille.

Description. A bed of fine-grained, light buff-weathering dolomite occurs near the top of the Cabot Head shale division of the Cataract formation in the vicinity of Dyer bay and Cabot head. At the type locality along the shore 3 miles west of Cabot head, on lots 3 and 4, concession XIV, St. Edmund township, Bruce peninsula, the dolomite is 8 feet thick and tends to be massive. *Favosites cristatus*, a beautiful branching coral, is its characteristic fossil and *Atrypa parksi* occurs rarely. Four feet of soft green shale intervenes between the top of the St. Edmund and the base of the Lockport dolomite, which is thin-bedded as exposed at Cabot head. The character of the St. Edmund at Cabot head is the same as at the type locality. At Dyer bay, the St. Edmund is poorly exposed, but is represented by slabs full of *Favosites cristatus*. The thickness appears to be less at Dyer bay, however, as is also the thickness of the thin beds at the base of the Lockport, judging from the covered interval. Southward on Bruce peninsula, the horizon of the St. Edmund dolomite is general talus covered. On Manitoulin island this dolomite has not been recognize and appears to be absent.

Microscopic Characters. Examined under the microscope, a specime of St. Edmund dolomite from the Clay cliff near Cabot head proves to l made up of an aggregate of small, uniform, rounded grains of dolomi with an average diameter of 0·01 mm.

A specimen of ferruginous sandy dolomite from near the top of tl t'ahot Head shale, at Limehouse, under the microscope proves to be con posed of rounded, worn dolomite grains from 0·09 to 0·22 mm. in diamete with a little ferruginous matter. The smaller dolomite grains interloc and show no signs of wear. Numerous sections of bryozoa are present.

.auna of St. Edmund Dolomite. This is recognized only at the e: posures in the vicinity of Cabot head, where it contains *Favosites cristat* Edwards and Haime and *Atrypa parksi* Williams. n. sp. (see Plates I\ V, and VII).

Origin of Cabot Head Member.

Of the origin of the Cabot Head shales Grabau says[1]: " Westwar however, the northern series of deposits (the Upper Juniata) were wash into the sea, the shore of which during Medina time did not probabl extend much farther east than Rochester. In western Ontario, encroacl ment of the sea began earlier, for here we have the marine Keppel (Man toulin, author) dolomite resting on a surface of Queenston. The dolomites were probably formed while erosion was beginning in the ea: and while the early Medina sands were spread by the streams over the ol Ordovicic land surface. Then as the product of this erosion was carri farther westward by repeated reworking, it finally entered the sea, whic had then advanced farther east. Thus the fossiliferous Medina beds western New York and the red and green Cabots Head sands which re on the Keppel dolomite were deposited."

The lack of sandstone in the north is doubtless due to the distanc from the shore and the source of the sedimentary materials. The fac however, that in the north the lower beds are red and in the south tl upper beds are red is not easily explained on Grabau's hypothesis alon particularly as there appears to have been continuous deposition of or kind or another throughout the region. However, highly oxidized sed ments on being washed out to sea met deoxidizing agents in the form plant and animal organisms, and it is clear that normal marine conditio extended farther eastward into New York during early Cabot Head tin than during late t'ahot Head and Grimsby time. Thus, in the Niaga peninsula the lower Cabot Head, and the underlying Manitoulin be clearly represent marine conditions and the colour is grey; higher be are sandy, representing shallower water, and red is the predominatii colour. Grabau derives the sediments forming these deposits from ti older, highly oxidized formations such as the Juniata-Bays. In tl Georgian Bay region, however, it is hardly probable that the origin of tl sediments is as simple as supposed by Grabau. The author has show

[1] Bull. Geol. Soc. of Am., vol. XXIV, No. 3, Sept., 1913, p. 469.
[2] Geol. Surv., Can., Mus. Bull. 45, p. 9.

that in the Lake Timiskaming region, a long period of erosion laid bare early Trenton fossils resting in an irregular basal conglomerate on the Pre-Cambrian rocks, and upon these deposits Niagara limestones were laid down. Elsewhere in the same region, as shown by G. S. Hume,[1] 70 feet of strata lie between the top of the Trenton and the base of the Niagara. Thus a considerable amount of erosion is shown to have occurred in the north during the period including the deposition of the Cabot Head shales. That some of the eroded material should be carried southward is altogether probable and the mingling of the detritus eroded from the Richmond shales with the materials coming from the Adirondacks and the nearby crystalline rocks, together with the deoxidizing influence of marine conditions, may easily account for the grey colour of the upper part of the Cabot Head shales. The red colour of the lower Cabot Head shales of the Georgian Bay region may be due to the inwashing of highly oxidized sediments from the nearby Pre-Cambrian areas to the east and north during the predominance of littoral conditions. The probability of sediments coming from these nearby sources cannot be ignored.

The St. Edmund dolomite (a magnesian calcilutite) shows evidence of being composed of dolomite flour and may with the Dyer Bay dolomite have been derived from older dolomites of the northern realm.

GRIMSBY SANDSTONE MEMBER.

Characteristics and Distribution.

The 50 feet of red and grey shale and sandstone at Niagara river, limited below by the Cabot Head shale and above by the Thorold sandstone, has been named by the author the Grimsby[2] member. At Grimsby, the type section consists of 12 feet of thick-bedded, mottled, red and grey sandstone, and 6 feet of grey shale which lies immediately beneath the Thorold sandstone. The underlying strata are red Cabot Head shales.

The Grimsby sandstones show considerable variation, some beds, particularly in the lower half, being decidedly shaly. Crossbedding is common in the thicker beds and lenses and pillow structure are fairly common. The pillow structure appears to the writer to be due to the rolling by current action of masses of sand over thin beds of mud, the incoming sand preserving the form. Current action was certainly present as shown by crossbedding.

The Grimsby sandstone is seen as far west as Dundas, where it is not easily differentiated from the Cabot Head shales, but is about 14 feet thick. Eastward, it increases in thickness, being as above stated, 50 feet at the American line. In Canada this member rests upon the Cabot Head shale, and is doubtless the inshore equivalent of the upper beds of that shale as seen north of Dundas. As far west as Stoney Creek the Grimsby sandstone is overlain by the Thorold quartzite. Whether the upper grey shales and sandstone at Hamilton are to be included with the Grimsby or the Thorold, it is not easy to decide, but on the evidence of pillow structure which is a well known characteristic of the Grimsby sandstone, the lower beds are referred to that member. It is pro-

Geol. Surv., Can., Sum. Rept., 1914, p. 191.
Geol. Surv., Can., Sum. Rept., 1913, p. 184.

able, that the upper, thin, grey sandstones and shales are a facies of Thorold sandstone.

On the evidence of well borings, 25 feet of sandstones at Port Bur- and 45 feet at Paris are referred to the Grimsby member. About 7 of sandstones and shales are referred to the Grimsby at Dundas, and Kelso, Grimsby beds are lacking. A well record at Beachville indic that the Grimsby sandstone is lacking there, and it is also lacking at l don. The northwestern, underground extension of the Grimsby app from the above evidence to be approximately limited by a line dr from the Niagara escarpment between Dundas and Kelso, southw passing to the north of Paris and to the east of Beachville and termina at the lake Erie shore perhaps in the vicinity of Port Bruce.

Microscopic Characters.

A specimen of green-grey sandstone, from a bed 9 feet above the l of the Grimsby member at Niagara river, proves under the microsc to consist of subangular to rounded quartz grains with an average diam of about 0·1 mm., some chlorite and kaolin, and a very little calcite.

Origin of Grimsby Sandstone.

As already described, Grabau's hypothesis[1] derives the Medina l from folded Juniata-Bays and Bald Eagle formations in the Appalacl region. Erosion having cut through the Juniata-Bays red beds, we next attack the grey Bald Eagle conglomerate. The Grimsby sandst appears to be an inshore phase of the early erosion of the Bald Ea The finer sediments were deposited farther out to sea, forming the up Cabot Head beds, and the coarser materials, mixed with the coarser hig oxidized materials from the red beds, were deposited in an estuary along the littoral zone of the sea in what is now the Niagara penin of Ontario. (For fossils see the following table and Plate VIII.)

Fauna of Grimsby Sandstone.

Genera and species.	Niag-ara river.	Thor-old.	Jor-dan.	Ham-ilton.	Water-down.	Flam-bor-ough tp.
Brachiopoda.						
Lingula cuneata Conrad	x	x	x			
Lingula clintoni Var "n...				?		
Dalmanella eugeni- Williams n. sp				x		
Camarotoechia (Ste ynchus) neglecta Hall.	x			x		
Pelecypoda.						
Pterinea cf undata (Hall ...	x			x	x	x
Pterinea brisa Hall ...						
Modiolopsis primigenia (Conrad)...	x[2]			x		
Modiolopsis orthonota (Conrad)	x[2]					
Modiolopsis kelsoensis Williams n. sp						x
Ctenodonta machæriformis (Hall...						x
Nuculites cf ferrugineum Foerste					..	x
Gastropoda.						
Bucanella trilobata (Conrad)....	x					

[1] Ibid, page 469
[2] Reported by Grabau

THOROLD SANDSTONE MEMBER.

Characteristics and Distribution.

The grey sandstone at the top of the Medina in New York, and southern Ontario, formerly known as the "Grey Band", has been called by Grabau[?] the "Thorold quartzite". The sandstone is light grey, thin to thick-bedded, and at Thorold is a true quartzite. Elsewhere, "sandstone" more fittingly describes it. The sandstone is commonly crossbedded and contains at Grimsby and elsew[he]re *Daedalus archimedes* Ringueberg and *Arthrophycus harlani* Hall.

At Niagara gorge the Thorold sandstone is 7 feet thick; it attains a thickness of about 12 feet near Decew falls, is 6 feet thick at Stoney Creek, and at Hamilton and northward is not clearly differentiated, if present, from the Grimsby sandstone and shale. At its lower contact the Thorold sandstone is marked by little more than a lithologic change from the Grimsby sandstone below. The top bed of sandstone upon which the Clinton [Reynales (Wolcott)] dolomite rests, is thick and free from calcareous matter. At some localities, however, the Reynales dolomite is arenaceous at the base.

Fauna of Thorold Sandstone.

Genera and species.	Niagara river.	Thorold.	Grimsby.	Dundas.
Worm burrows.				
Daedalus archimedes (Ringueberg).	:		x	x
Arthrophycus alleghaniensis (Harlan).	x	x	x	x

Microscopic Characters.

A specimen of Thorold quartzite from Thorold, Ont., examined under the microscope, proves to consist of rounded to subangular quartz grains of 0.10 mm. average diameter, closely interlocked. Along their contacts, recrystallization has taken place, and some secondary silica has been deposited.

Underground Extension.

As stated above, the Thorold sandstone extends along the Niagara escarpment as far west as the vicinity of Hamilton. Well records indicate that its extent southward is not great, only red sandstones being mentioned in well logs, excepting at Port Colborne, where thin-bedded, white sand-stones occur interbedded with red shales, and at Welland where grey shales are recorded above the red sandstones. It appears fairly clear that the Thorold sandstone does not extend underground much beyond a line drawn from Port Colborne northwest to Hamilton.

Origin.

The origin of the Thorold sandstone has been explained by Grabau as the result of the erosion of the Bald Eagle conglomerate, after the Juniata-Bays red beds had been cut through to furnish material for the formation of earlier Medina members. The Thorold as well as the Grimsby sand may, however, have been derived from the crystalline rocks of the Appalachian after the removal of the younger strata.

The origin of the Medina sediments, which are at Niagara river, in ascending order, sandstones, shales, sandy dolomite, shales, and sandstones, must be considered from the standpoint of the relative position of the sea and land as well as from the standpoint of the source of sediments

A. W. G. Wilson[1] has shown that when the rate of depression of the land is greater than the rate of supply of detritus "there will be a uniform shoreward overlap of the zones of deposition", that is from deep to shallow water, limestones will overlap shales, which in their turn will overlap sandstones and conglomerates. Should the rate of depression of the land, later become less than the rate of supply of detritus, the overlap would be reversed, conglomerates and sandstones overlapping shales and shales overlapping limestones to seaward. Thus with a reversed cycle, limestones would be enclosed in shales, and shales in sandstones in a direction from sea to land. It will be readily seen from the sections of the Medina members that this is exactly the same as in the case of the Medina. In the first part of the cycle of sedimentation, whirlpool sandstone was overlapped landward by shales which comprise the base of the Manitoulin beds in its southern extension, and the shales and sandstones both were overlapped by dolomite. With the reversed order Cabot Head shales overlapped Manitoulin dolomite to seaward, and Grimsby sandstones overlapped the shales in the same direction. The Thorold sandstone and the Oneida conglomerate of New York state may represent a temporary mastery of the depression of the land over the supply of detritus. That is, the shaly character of the upper Grimsby sandstone, followed by the limited deposit of Thorold sandstone, may well indicate a slight advance of the sea at the end of Medina time.

The reason for the change of balance in the factors controlling sedimentation, as applied to Medina sediments, is to be found on the one hand in the relative change in elevation of land and sea, and on the other in the varying sources of supply of sediments as erosion progressed

CORRELATION AND AGE OF THE MEDINA-CATARACT FORMATION.

As already stated the Medina is represented in eastern New York state entirely by sandstone and conglomerate. This is because of proximity of the Appalachian mountains which were the main sources of the sediments. Bassler[2] correlates the Oneida-Shawangunk-Green Pond formation of central and eastern New York state with the Medina below the Thorold

According to Ulrich[3] the Clinch sandstone of Tennessee is probably the equivalent of the Tuscarora sandstone of Pennsylvania and in general also the equivalent of the Medina (restricted use) of New York and Ontario

Can. Rec. of Sc., vol. IX. No. 2, 1903, p. 119.
U. S. Nat. Mus., Bull., 92, vol., I and II, 1915, pl. 3.
Bull Geol. Soc. Am. No. 7 Vol XXII, 1911, p. 468

Savage has recently described in detail the stratigraphy and palæontology of the Alexandrian series in Illinois and Missouri.[1] He has divided the series in ascending order[2] into Girardeau limestone, Edgewood limestone, Essex limestone, and Sexton Creek (Brassfield) limestone.

In a later article,[3] Savage indicates important correlations. After referring *Conchidium decussatum* Whiteaves to the genus *Virgiana* recently described by Twenhofel, Savage says in part as follows: "In the Grand Rapids region the layers containing *Virgiana decussata* are succeeded by strata which contain the fossils *Pterinea occidentalis*, *Isochilina grandis* var. *latimarginata*, and *Leperditia hisingeri* var. *fabulina*. In the Hudson Bay region a zone a few feet above the horizon of *Virgiana decussata* furnished shells of *Camarotœchia? winiskensis*, *Pterinea occidentalis*, *Isochilina grandis* var. *latimarginata*, and *Leperditia hisingeri* var. *fabulina*. In the northern peninsula of Michigan early Silurian strata containing *Camarotœchia? winiskensis*, *Isochilina grandis* var. *latimarginata*, and *Leperditia hisingeri* var. *fabulina* overlie the strata containing *Virgiana mayvillensis*, which is a near relative of *Virgiana decussata*. In eastern Wisconsin *Virgiana mayvillensis* occurs in the uppermost layers of the Mayville limestone, above which there is a stratigraphic break, the horizon of *Camarotœchia? winiskensis*, *Isochilina grandis* var. *latimarginata*, and *Leperditia hisingeri* var. *fabulina*, present farther east in northern Michigan, having been removed by erosion. However, there is no doubt that the strata which in northern Michigan contain *Virgiana mayvillensis* correspond in age to those containing the same species in the upper part of Mayville limestone in Wisconsin, as they are clearly a northeastward continuation of the same beds"......

"In Wisconsin there was found in the quarry near Pebbles a zone only a few feet below the horizon of *Virgiana mayvillensis* and apparently conformable with it, which yielded such characteristic Edgewood species of fossils as *Dalmanella edgewoodensis*, *Rhynchonella? janea*, *R¹ jncholreta parea*, and *Atrypa putilla*. The position of *Virgiana mayvillensis* in Wisconsin in the upper part of the Mayville limestone, which at a slightly lower level contains a characteristic Edgewood fauna, indicates that this horizon is Alexandrian (late Edgewood) in age. It is also significant that the strata containing *Virgiana mayvillensis* in Wisconsin seems to occupy about the same position in the Silurian column as do the strata which contain *Virgiana barrandei* in the Beesie River formation of Anticosti island."

"In Michigan the strata containing *Camarotœchia? winiskensis*, *Isochilina grandis* var. *latimarginata*, and *Leperditia hisingeri* var. *fabulina* conformably overlie the *Virgiana mayvillensis* beds, and thus are thought to correspond in age to about that of the Sexton Creek or Kankakee limestone which overlies the Edgewood in Illinois and Missouri, but they were deposited in a different geologic province."

"The close correspondence in the fauna of the strata overlying the *Virgiana mayvillensis* zone in northern Michigan with that of the strata above the horizon of *Virgiana decussata* in the Hudson Bay and Saskatchewan regions leaves no doubt of the equivalence of the strata containing

[1] Illinois Geol. Surv., Bull. 23, 1917, pp. 67-199.
[2] Ibid. p. 73.
[3] Correlation of the early Silurian rocks in the Hudson Bay region, Jour. of Geol., vol. XXVI, 1918, pp. 336-340.

this fauna in the areas above mentioned. They also prove that the *Virgia mayvillensis* zone in Wisconsin and Michigan, and the *Virgiana decuss* zone in the Hudson Bay and Saskatchewan localities represent the sa stratigraphic horizon."

"Besides the above-mentioned localities Hume[1] has found ea Silurian strata containing *Camarotœchia? winiskensis*, and numere ostracods in the lake Timiskaming area that he correlates with the Catara formation, which doubtless corresponds with the *Camarotœchia? winisk sis, Isochilina,* and *Leperditia* horizon in the regions above described."

"Kindle[2] found Silurian strata several hundred miles north of t Grand Rapids locality, in the vicinity of The Pas from which he obtain the fossils *Camarotœchia? winiskensis, Plerinea* cf. *occidentalis,* and *Le p ditia* cf. *hisingeri.* This fauna also indicates a horizon about equival to that of the Silurian in the lake Timiskaming region and to the stra containing *Plerinea occidentalis, Isochilina,* and *Leperditia,* above t *Virgiana decussata* horizon in the Grand Rapids section, the latter horiz not being exposed in the more northern locality. From the similarity the faunas of the *Virgiana* zone, and of the higher strata containing *Ca arotœchia? winiskensis, Isochilina grandis* var. *latimarginata,* and *Leperd hisingeri* var. *fabulina* in the regions above described it is inferred th during the time those strata were laid down the above-mentioned regio were a part of the same province or basin of deposition, which w rather broadly connected northward with the Arctic ocean."

The presence of *Virgiana mayvillensis* in the Dyer Bay dolomite clea establishes its equivalence with the *Virgiana* zones described by Sava As the Dyer Bay dolomite which is near the top of the Cataract format is of upper Edgewood age, it seems probable that the remainder of the Cab Head shale and the Manitoulin dolomite are of lower Edgewood age. the Girardeau and the Manitoulin dolomite have little in common.

The diagnostic species common to the Manitoulin dolomite and t divisions of the Edgewood limestone indicated are:

Rhipidomella hybrida Channahon limestone, Ill.
Rhynchonella? janea Edgewood limestone, Ill.
Bucania cf *exigua* Edgewood limestone, Ill.

The species common to the Manitoulin dolomite and Brassfie limestone are:

Clathrodictyon vesiculosum Sexton Creek limestone, Ill.
Homotrypa? confluens Brassfield limestone, Dayton, Ohio.
Pachydictya turgida Brassfield limestone, Dayton and Fair Haven, Ohio.
Strophonella striata Brassfield limestone, Ky.
Herbertella fausta Brassfield limestone, Ohio.
Rhipidomella hybrida Brassfield limestone, Ky.

The species common to the Dyer Bay dolomite and the Edgewo limestone are:

Schuchertella curcistriata Channahon limestone, Ill.
Rhynchotreta thebesensis var. *multistriata* Edgewood limestone, Ill.
Rhynchonella? janea Edgewood, Ill. and Mo.

It will be seen that the fossil evidence is not definite enough to ma close divisional correlations. An anomaly is present in the close relati

[1] G. S. Hume, "Palæozoic rocks of lake Timiskaming area," Geol. Surv., Can., Sum. Rep., 1916, pp. 18 Fossils reported by Charles Schuchert in a personal letter.
[2] E. M. Kindle, op. cit., p 17.

ships of the faunas of the Manitoulin dolomite and the Brassfield limestone, for the Dyer Bay dolomite, which is much higher than the Manitoulin dolomite, is more closely related by its fauna to the Edgewood
limestone which underlies the Brassfield limestone.

Schuchert's summary of the relation of the Cataract formation to other
formations, written before the completion of the work done by Savage and
the author, is given below for comparison. After discussing the relation
of the Cataract and Medina faunas, Schuchert adds:

"It should also be stated that the Medina is of the Appalachian
province, while the Cataract is either of the St. Lawrence or of the Arctic
realm. These waters came in over the continent from different oceanic
areas and accordingly have different organic associations. Hence the
similarity of the biotas cannot be close, and this, with the marked difference
in sedimentation, gives additional reason why considerable weight is laid
on the forms held in common, as showing that both formations are the
deposits of about one time."

Under "Relation to the Brassfield," Schuchert says: "The Cataract
may also be compared with the Brassfield formation of Ohio and Indiana,
as the two are clearly related, and also, as both are of a limestone facies.
Between the two there are twenty-four forms in common, and of these
the following have the most significance in correlation. *Clathrodictyon
vesiculosum, Areracularia (?) gracilis* (in Ohio, *A. clodonensis*), *Rhinopora
verrucosa, Phaenopora ensiformis, Callopora vaughnpora, Homotrypa* (?) *con-
tiguus*, and *Hebertella fausta*.

Under "Relation to the Siluric of Anticosti," Schuchert says: "The
Cataract does not readily correlate with the Anticosti section because of
the marked differences and generalized character of the faunas there, and
more especially because of the long range of most of the species. With
the Beesie River, the only guide fossil in common is *Catospira planoconvexa*
appearing about 70 feet above the base of the Beesie River); but the Anticosti individuals are only half grown compared with those of the Cataract,
a condition seemingly in harmony with the conclusion that the latter are
of a younger time. Then the absence in the Cataract of the Beesie River
guide, *Clorinda barrandei*, also seems to indicate that the former formation
is of younger age. At the top of the Beesie River, however, the fauna is
more like that of the Cataract, and this similarity continues in the succeeding 300 feet of the Gun River formation."

Schuchert[2] says "Even though the Medina, Cataract, and Brassfield
are correlates of one another, it does not follow that each one is wholly
the equivalent of any other. Each formation invades eastern North
America from a different direction and each one has its own peculiar
faunal assemblage. They, therefore, represent three physical provinces
and marine basins. The Medina is of the northern Appalachian province,
is a sandstone formation, and finally invades to a slight extent the area
of the Cataract. The Brassfield province lies, in the main, west of the Cincinnati axis, is of southern origin, with limestone-making seas, spreads
also up the southern portion of the Appalachian province, and finally
evades slightly the area of the Cataract sea. On the other hand, the
Cataract province spreads westward through the Saint Lawrence em-

Bull. Geol. Soc. Am., vol. XXV, 1914 p. 294
See page 294

bayment, and finally, in eastern Ontario and northeastern Ohio (kn from the Clinton oil wells), unites with the other two provinces; bu the Medina waters form a shoal sandy area in northeastern Ohio betw the other two provinces; very few of the species of either area intermig Probably it would be more correct to state that the normal marine junc of the Cataract and Brassfield seas is prevented by the Medina d For these reasons Medina, Cataract, and Brassfield are to be retaine names for independent marine faunas and formations."

According to Twenhofel[1] the Cataract formation is probably t correlated with the lower portion of the Gun River formation of Antic island. No sediments in the Silurian section at Arisaig have been ident as being as old as the Medina-Cataract, although it is possible that base of the Beechhill Cove formation may be of this age.

Bassler correlates the Medina-Cataract with the Lower Llando of Wales and Shropshire and the Lower Birkhill of Scotland.

CLINTON FORMATION.

As here described the Clinton formation includes the shales limestones between the Medina-Cataract formation below and the Roc ter[2] formation above.

DEFINITION AND GENERAL DESCRIPTION.

As already stated, the Manitoulin dolomite of Ontario, along the overlying shales, were by Sir William Logan misnamed the Clin formation: This was partly due to the mistaken idea that the Whirl sandstone of Ontario was the "Grey band" of New York, and pa because the Manitoulin dolomite was a convenient horizon-marker. quote Logan[3], "In Canada, for reasons which will be stated in descri the Niagara formation, it is found convenient to limit the Clinton to strata beneath the Pentamerus band, and to include this band in Niagara formation. On the Niagara river, the Clinton is thus lim to a few feet, but it gradually augments to the northward". And u "Niagara formation" Logan[4] says: "We therefore, propose to includ the Niagara series, the two bands of limestone which underlie the sh and which, in New York, constitute the upper part of the Clinton for tion. So far as they have been examined in Canada, these two l stone bands contain no Clinton fossils, but such as pass upwards into Niagara; and the upper band here possesses one or two species, wl in New York, are considered to belong to the latter group only. T would thus appear, at present, to be no palæontological reason why t limestones may not be considered the base of the Niagara format

[1] Geol. Surv., Can., Mus. Bull. 3, 1914, p. 29
[2] It is now the custom of the United States Geological Survey and the New York State Survey to inclu Rochester shale in the Clinton formation, on the ground that the Rochester shale is included in the Clinton ser Clinton, New York. In the opinion of the Canadian Geological Survey, this contention is not proved. other hand, in western New York and in Ontario, the Rochester was clearly considered a part of the Niaga mation by Hall, Logan, and other early workers in the region. There is no denying the continuity of sedimer and faunal development from the Irondequoit to the Rochester, and according to Chadwick, what has considered by the author only an apparent break, at the top of the Rochester, is due to disconformity an Lo A number of fossils, however, are common to the Rochester and the Lockport and for the present the other the term Clinton is maintained by the Geological Survey, Canada.
[3] Geology of Canada, 1863, p. 312.
[4] Ibid, page 322.

while geographically they present a very marked feature for a considerable distance, and afford a convenient means of describing the distribution of the two formations."

Admitting the close relationship between the Clinton and Niagara formations, it is nevertheless impossible to ignore the distinct characters both palæontologic and stratigraphic of the Clinton formation, even in its vanishing condition near Hamilton.

Only the upper part of the Clinton of New York is present in Ontario. Thus of the seven divisions recognized by Chadwick[1] at Rochester, only three occur with certainty at Niagara, and a fourth is doubtfully recognized. The divisions at Rochester are from the base up: Maplewood (Sodus) shale; Bear Creek limestone and shale; Furnaceville iron ore and limestone; Reynales (Wolcott) limestone; Sodus and Williamson (Williamson) shales; and Irondequoit limestone. At Niagara river, a shale probably of the age of a part of the Furnaceville beds (formerly called the Sodus shale), the Reynales (Wolcott) dolomite, and the Irondequoit dolomite are well represented and the Williamson shale is suggested by the shaly parting between the Reynales and Irondequoit. The total thickness of the Clinton at Rochester is about 80 feet, and at Niagara river 23 feet: Hall called the Clinton the Protean group because of its varied character, shale being dominant in eastern New York, and shale and calcareous beds being about equally represented in western New York. The variation between Rochester and Niagara river is indicative of the protean character of the Clinton and the changes are striking from place to place in Ontario. The Clinton outcrops at many places along the Niagara escarpment between Niagara river and Kelso.

The Furnaceville shale is not seen west of Niagara river, unless a few inches of shale at DeCew falls represents it. The Irondequoit dolomite thins out to the westward, being last seen along with the overlying Rochester shale near Waterdown. The Reynales dolomite is 6 feet thick at Kelso and contains *Pentamerus oblongus* near the base, just as at Thorold and other more eastern localities, but it has not been identified farther west or north. A similar bed at Limehouse contains *Dictyonema retiforme*, and probably represents the horizon of the DeCew water-lime rather than the Reynales dolomite. Well records indicate that Clinton dolomite extends westward beneath younger formations as far as lake St. Clair where about 5 feet of limestone occurs at the base of the Rochester shale. Samples of drillings from a well on lot 1, concession V, Bayham township, Elgin county, and from another at Beachville, Oxford county, indicate 20 feet or more of dolomite at the Clinton horizon. For the distribution of the Clinton, see Maps 1714 and 1715. For fossils see tables under description of members and Plates IX and X.

THE FURNACEVILLE (SODUS) SHALE MEMBER.

Description.

The Furnaceville (Sodus) or lowest division of the Clinton at Niagara gorge is a soft, fissile, bluish-grey shale weathering greenish. It rests

---

[1] The nomenclature and correlation of the Clinton formation of New York state has been recently revised by Chadwick. His nomenclature is here followed. The terminology formerly recognized is placed in brackets.

evenly and sharply on the Thorold sandstone, with no indication of transition beds. The best outcrops are along the New York Central and Hudson River railway between Niagara Falls and Lewiston and about seven-eighths of a mile north of Niagara university; here the shale is 4 feet thick. A few inches of green shale occupies the position of the Furnaceville at Decew falls but it is unknown elsewhere in Ontario. The accompanying table gives the Furnaceville fossils found on the Niagara river.

The Furnaceville (Sodus) Shale Fauna of the Niagara River

Rhafistrophora wilsa (Hall) [G]
Rhynchotrema cuneata americana Hall
Calospira hemispherica (Sowerby) G.
Plectambonites corrugata Conrad[?]
Corunlonta cllipticus (Hall)[?]
Coraxinugc decussa Whitfield and Hovey.
Orthis sp. [2 sp]

Of the above, *Calospira hemispherica*, according to Chadwick, is found only in the Furnaceville and Sodus divisions.

THE REYNALES (WOLCOTT) DOLOMITE MEMBER.

Description.

The Reynales division of the Clinton is a dark grey, fine-grained dolomitic limestone, approaching a true dolomite in composition (see chemical analysis, page 110). It rests with sharp even contact on the Furnaceville shale at Niagara gorge, forms the basal Clinton member along the Niagara escarpment between Thorold and Waterdown, and is the only representative of the Clinton from the vicinity of Waterdown to Kelso, beyond which it has not been identified. West of Niagara river the Reynales dolomite rests successively on Thorold sandstone and Cabot Head shale. The contact is everywhere sharp, but the basal dolomite is arenaceous at some localities.

Good exposures ... occur along the New York Central and Hudson River rail... Niagara Falls and Lewiston, from one-half to seven... rth of Niagara university; at Niagara Glen on the Ont... er; and at numerous localities along the Niagara escar... t. Some of the best localities for study are at Dec... by Stoney Creek, Hamilton, Dundas and Kelso. At Niaga... Reynales is 12 feet thick and weathers into thin beds. West... aner and is commonly thick-bedded At Jordan, 3 feet of shale... near the middle of this division.

Guide Fossils.

Hyattidina congesta and *Stricklandinia canadensis* are guide fossils to the Reynales dolomite. *Pentamerus oblongus*, formerly considered guide fossil for New York state, is shown by Chadwick to occur at

[These reported by Grabau are marked "G"]
[Chadwick identifies this as *Coelospira intrens* (Vanuxem)]

n the Wa... limestone, used in his revised sense. At Niagara gorge,
P. oblongus has not been reported. As will be seen on page 1.. however,
Pentamerus oblongus has a wide range. The other Reynales fossils with
the ... distribution ... work ... out by the author are given in ... full ... tab...
...

Genera and species

	...	Buel	Lex ...	Grimsby	Hamilton	Lincoln	Index ...
Coral							
Cyathotrophis gracilis Hall							
Anthozoa							
Zaphrentis turbinata (Hall)							
Favosites niagarensis Hall							
Bryozoa							
Fenestella concentrica Hall	...						
Brachiopoda							
Strophodonta profunda Hall	...						
"Strophonella" patenta (Hall)	...						
Plectambonites transversalis (Wahlenberg)		...					
Schuchertella subplana (Hall)		...					
Platystrophia biforata (Schlotheim)							
Stricklandinia canadensis (Billings)							
Pentamerus oblongus Sowerby							
Camarotoechia (Stegerhynchus neglecta Hall	...	...		...	...		-
Spirifer radiatus (Sowerby)							
Hyattidina congesta (Conrad)	...						
Coelospira plicatula (Hall)	...						
Gastropoda							
Diaphorostoma niagarense Hall							
Pelecypoda							
...mus vertebratum (Hall)							
...ta							
Trochurus ornatus Hall and Whitfield							

... Reported by Grabau.
... Found by the author

WILLIAMSON SHALE MEMBER.

Description.

The Williamson shale is an important member of the Clinton formation at Rochester, having a thickness along with the underlying Soslus
shale of about 24 feet. At Niagara gorge, Grabau and Kindle have
reported that this shale is entirely absent. There is, however, at the
contact between the Reynales and the Irondequoit dolomites, about
1 inch of grey shaly weathering rock that suggests the attenuated edge

Schuchert criticizes this statement, stating that he has always considered the Clinton species des...
... the Lockport. The author has been unable to find distinctive characters, although the presen... ...
... at Perce

of the Williamson. At DeCew falls, 4 inches of definitely defined, d grey shale separates the Reynales and Irondequoit dolomites. It probable that this is the western remnant of the Williamson shale. fossils have been found in it.

IRONDEQUOIT DOLOMITE MEMBER.

Description.

The Irondequoit member of the Clinton is a light grey or buff, cr talline, dolomitic limestone, approaching a true dolomite in composit. (see chemical analysis, page 111). At Niagara gorge, reef structur consisting of lenses of dense, amorphous dolomite 6 to 35 feet across a as much as 10 feet thick, are common in the upper part of the Irondequ dolomite and in places extend upward into the overlying Rochester sh: These reefs are composed mostly of undetermined bryozoan remains a contain as well a rich brachiopod fauna which includes numerous spe common in the Rochester. Free swimming animals, such as the trilobit also frequented the openings in the reefs. Corals were practically abse with the exception of the little cup coral, *Enterolasma caliculus*. *Lichena concentrica* is the one identified species of bryozoa present.

The lower contact of the Irondequoit, whether with the Williams shale or the Reynales dolomite, is generally sharp and clear-cut. one quarry in Waterdown, however, the base of the Irondequoit is cement to the top of the Reynales. The Irondequoit is well exposed along w the Reynales dolomite as far west as Waterdown. At the Niagara go it is 7 feet thick and consists of a 1-foot bed at the base and an up massive 6-foot bed. At Waterdown the section consists of a basal l of argillaceous dolomite 1 foot thick and an upper bed of crinoidal dolom 3½ feet thick. Northward the Irondequoit has not been reported a it is known to be absent at Kelso.

Because of its thick bedding and toughness the Irondequoit is kno to the quarrymen of the Hamilton area as the " Nigger Head " bed a is commonly left as a quarry floor.

The Irondequoit fauna is closely related to the fauna of the overly Rochester shale and the reef structures represent transitional conditio The fauna with its distribution is shown in the following table.

[Kindle, E. M., U.S. Geol. Surv., Niagara Folio, No. 190, p. 7.]

Distribution of Irondequoit Fauna in Ontario.

Genera and species	Niagara river.		DeCew falls.
	Bedded formation.	Reefs.	Bedded formation.
Anthozoa.			
Enterolasma caliculum (Hall)	g		
Crinoidea.			
Glyptocrinus plumosus Hall			
Eucalyptocrinus crelatus levis Grabau and Shimer			
Bryozoa.			
Fistulella concentrica Hall			
Brachiopoda.			
Leptaena rhomboidalis (Wilckens)	xg	g	x
Stropheodonta profunda (Hall)	g		
Strophonella? patenta (Hall)	xg	g	
Plectambonites transversalis (Wahlenberg)		g	
Schuchertella subplana (Hall)	xg	g	
Orthis flabellites Foerste	g		
Dalmanella elegantula (Dalman)	x		
Rhipidomella hybrida (Sowerby)			
Anastrophia interplicata (Hall)			
Pentamerus oblongus Sowerby	g		
Clorinda fornicata (Hall)	xg		
Rhynchotreta cuneata americana (Hall)			
Rhynchotreta robusta (Hall)	v		
Camarotoechia (Stegerhynchus) neglecta (Hall)	g		
Camarotoechia? acinus (Hall)	g		
Atrypa reticularis (Linnaeus)	xg	g	
Atrypa nodostriata Hall	x	g	
Atrypa rugosa Hall		g	
Spirifer radiatus (Sowerby)	xg	g	v
Spirifer niagarensis (Conrad)	xg	g	
Spirifer crispus (Hisinger)		g	
Spirifer corallinensis (Grabau)	v		
Spirifer (Delthyris) sulcatus (Hisinger)		g	
Whitfieldella nitida (Hall)		g	
Whitfieldella nitida oblata (Hall)	xg	g	
Whitfieldella (intermedia) (Hall)		g	
Whitfieldella cylindrica (Hall)	xg		x
Pelecypoda.			
Pterinea brisa Hall			Hamilton?
Gastropoda.			
Diaphorostoma niagarense (Hall)		g	
Cephalopoda.			
Dawsonoceras americanum (Foord)			
Trilobita.			
Bumastis ioxus (Hall)	g		
Calymene niagarensis Hall	g		

Grabau reports one specimen from the Niagara gorge
Of the above, Rhynchotreta robusta (Hall), may be taken as a fair guide fossil to the Irondequoit dolomite
Found also in the Rochester shale
g = Reported by Grabau
x = Found by author

CORRELATION OF CLINTON MEMBERS.

The direct correlation of the Clinton formation of Ontario with th of New York state is evident from the terminology and descriptions giv above. According to Bassler[1], who made use of the latest reports his work, it is correlated with other formations as follows (the auth however, has eliminated formations correlated with the Rochester whi is included in the Clinton by Bassler): with the Clinton group of Pen sylvania, Maryland, and Virginia including the Cacapon member at t base and the Keefer sandstone near the top; the Dayton limestone a "Niagara" shale of Ohio; the Crab Orchard formation of eastern Ke tucky; the middle part of the St. Clair limestone of northern Arkans; the St. Clair limestone of central Oklahoma; the upper part of the G River, the Jupiter River (Williamson and Irondequoit), and a pa of the Chicotte (Irondequoit Rochester) formation of Anticosti[2] islan the Beechhill Cove and Ross Brook formations of Arisaig; and the Up Llandovery and Tarannon of Wales and Shropshire and the Upper Birkh and Tarannon of Scotland.

In the light of the recent revisions of the Clinton formation, Schuch informs the author that he still connects the Clinton sea of New Yo and Ontario with the gulf of Mexico, as in his "Paleogeography of Nor America."[3] There was a land barrier in eastern North America whi separated the equivalent seas of the gulf of St. Lawrence and the Ne York-Ontario areas. The deposits of New York-Ontario are those an oscillating sea in which the adjustments of elastic to calcareous se beds were nicely balanced.

NIAGARA GROUP.

Definition and Use of Terms.

The Niagara group[4] as defined in this report includes the Roches (Niagara) shale and the Lockport (Niagara) dolomite; or in other wo the strata between the Clinton dolomite and where this is lacking, Cataract formation below and the Guelph formation above.

The Rochester is the more important member of the Niagara central New York, but thins out westward, and in Ontario is confin in outcrop to the Niagara peninsula. The Lockport dolomite is clea

[1] U. S. Nat. Mus., Bull. 92, vol. II, 1915, pls. 3 and 4.

[2] Geol. Surv., Can., Mus. Bull. 3, 1914, pp. 20 and 21.

[3] Bull. Geol. Soc. Am., vol. XX, 1919, p. 45.

[4] This is the usage of Logan, who used "formation," "series," or "group" without distinction), Billings, Whites and all former Canadian geologists, who followed the terminology of the New York State survey. The us supported by the fact that many species of fossils, in spite of [illegible] pass from the Roches shale upward into the Lockport dolomite. As will be described under the [illegible] there is an appa break at the top of the Rochester, but the author has seen little field evidence that this is the result of more t current action at the commencement of Lockport time. Chadwick, however, has found evidence in New Y State indicating [illegible] and disconformity at the top of the Rochester. As recognized by all close student the Clinton, and recently emphasized by Chadwick, the Irondequoit and Rochester are so closely related faun that they may eventually be placed in a formation by themselves. This would separate the Rochester from [illegible] related lower Lockport. On the other hand, to place the closely related Irondequoit and Rochester separate division is decidedly preferable to including the Rochester with the Clinton formation, from the lo members of which both it and the Irondequoit differ materially. Until a satisfactory solution of the terms of the Rochester and Clinton is reached, the author prefers to retain the [illegible] recognized [illegible]

defined at Rochester and westward and northward it is of increasing thickness, being the great cliff-forming dolomite of the Niagara escarpment.

It is this Lockport dolomite and its equivalents under other names which have made the Niagara the best known of all the Silurian formations in the regions of the Great Lakes, lake Timiskaming, lake Winnipeg, Hudson bay, and the Arctic.

ROCHESTER SHALE FORMATION.

Description.

The Rochester shale takes its name from the type section at Rochester where according to Kindle[1] it is 85 feet thick. This, formerly known as the Niagara shale, consists of fissile, dark grey shale, containing calcareous beds. The harder or more calcareous beds generally occur near the top, but at Niagara falls they are most common near the middle and in the lower half of the shale.

The Rochester shale, as stated above, is closely related to the Irondequoit dolomite upon which it rests and along with which it thins out to the north of Waterdown. The contact is a gradation from shaly dolomite to shale and at Niagara river amorphous reefs project upward from the Irondequoit into the Rochester. The fauna is likewise transitional, the species of the upper shaly Irondequoit beds and of the reefs, being commoner in the Rochester than in the Irondequoit below.

The fauna of the Rochester shale is rich in genera and species, and with the exception of the *Eurypterida* has all the main Silurian classes of marine organisms well represented. The greatest profusion of species, however, is to be found among the brachiopods, bryozoa, and trilobites. For the fauna and its distribution, see the accompanying table and Plates X, XII, and XIII.

[1] Niagara Folio No. 190, p. 7.

Distribution of Rochester Fauna in Ot.

Genera and species.	Niagara river.		Thorold	
	Lower.	Upper.	Lower.	Upper
Plants.				
Euthotrephis gracilis. Hall			x	
Radiata.				
Dictyonema retiforme (Hall)	x		x	
Dictyonema gracile Hall				
Anthozoa.				
Enterolasma caliculum (Hall)	x		x	
Zaphrentis turbinata (Hall)	x		x	
Favosites hisingeri Edwards and Haime	x		x	
Favosites niagarensis Hall	x			
Cladopora seriata Hall	x			
Stratopora flexuosa Hall				
Cystoidea.				
Caryocrinites ornatus Say	x		x	
Crinoidea.				
Stephanocrinus angulatus Conrad	x		x	
Eucalyptocrinus crelatus laevis (Graham and Shimer)	x			x
Ichthyocrinus laevis Conrad	x			
Annelida.				
Cornulites bellistriata Hall	x			
Bryozoa.				
Chilotrypa ostiolata (Hall)	x		x	
Batostomella granulifera (Hall)	x		x	
Nicholsonella florida (Hall)	x		x	
Trematopora tuberculosa Hall	x			
Hallopora elegantula (Hall)	x			
Chaetopora seperatostricta Hall	x			
Stomatopora diffusa (Hall)	x			
Lecnotella elegans Hall	x			
Semicoscinium tenuiceps (Hall)	x			
Polypora incepta Hall	x		x	
Lithopora alpiccornis Hall	x			
Pinnatopora dichotoma Hall	x			
Fenestella conventricta Hall	x		x	

Lingula (lamellata) Vanuxem
Lingula (clintoni) Hall
Leiolepis spinulosus Hall
Dictyonella coralliifera (Hall)
Leptaena rhomboidalis (Wilckens)
Strophomella striata (Hall)
Strophomella? patenta (Hall)
(Strophomella?) deweenensis Williams
Plectambonites transversalis (Wahlenberg)
Schuchertella subplana (Hall)
Orthis flabellites Foerste
Schizophoria punctostriata Hall
Orthostrophia (Schizogramma) fasciata Hall
Delthonella elegantula (Dalman)
Rhipidomella hybrida (Sowerby)
Anastrophia interplicata (Hall)
Rhynchotreta cuneata americana (Hall)
Rhynchotreta robusta (Hall)
Camarotoechia obtusiplicata (Hall)
Camarotoechia (Stegorhynchus) neglecta Hall
Atrypa reticularis (Linnaeus)
Atrypa nodostriata Hall
Atrypa rugosa Hall
Spirifer radiatus (Sowerby)
Spirifer niagarensis (Conrad)
Spirifer crispus (Hisinger)
Spirifer corallinensis (Grabau)
Spirifer (Delthyris) sulcatus (Hisinger)
Trematospira camura Hall
Whitfieldella nitida (Hall)
Whitfieldella nitida oblata (Hall)
Whitfieldella intermedia (Hall)
Whitfieldella cylindrica (Hall)

Pelecypoda.
Pterinea emacerata (Conrad)
Pterinea undata (Hall)
Leiopteria subplana Hall

Gastropoda.
Platyceras angulatum (Hall)
Diaphorostoma niagarense (Hall)
Diaphorostoma hermaphroditum (Hall)

Cephalopoda.
Trochoceras niagarense Hall
Trematodiscus minutus Hall

Found only in the Rochester e-shale
found only in the Rochester

Distribution of Rochester Fauna in Ontario— Co

Genera and species.	Niagara river.		Thorold.		D
	Lower.	Upper.	Lower.	Upper.	
Cephalopoda.					
Orthoceras uniomense Worthen					
Kionoceras cancellatum Hall					
Ostracoda.					
Aechmina spinosa (Hall)	x²	x			
Trilobita.					
?Homalonotus delphinocephalus (Green)	x	x			
Bumastis ioxus (Hall)	x	x			
?Dalmanites limulurus (Green)	x	x			
Calymene niagarensis (Hall)	x				
Arctinurus boltoni (Bigsby)	x				
Encrinurus ornatus Hall and Whitfield					

NOTE: Fossils listed by Graham are marked "g," those found by the author are marked "x," those reported by Grabau, at which found
1 Found only in the Rochester shale
2 Good guides to the Rochester

LOCKPORT DOLOMITE FORMATION.[1]

General Description.

The Lockport dolomite is named from the type section at Lockport, New York, where the section consists "chiefly of dark grey to chocolate coloured dolomite".[2] Two horizons of Guelph fossils known as the lower and upper Shelby have been found in New York with an intervening zone of Lockport fossils. This has led to considerable confusion, but Chadwick and the author have identified the Eramosa beds (at the top of the Lockport) at Niagara river and at Rochester, thus showing that the Guelph formation is well defined in New York state and includes the beds (so far as identifications have been made) from the base of the upper Shelby up. At Niagara river the crest of the American falls is at the top of the Eramosa beds which are also exposed along the Gorge route below the power-house on the New York side. They are about 12 feet thick and comprise Clarke and Ruedemann's division 4,[3] which contains a "continuous bed of chert nodules near the top", thought to be the "probable horizon of the upper Shelby fauna." Although the fossil occurrences do not exactly coincide, the conditions are similar to those described below for the Eramosa beds near the city of Guelph, where some Guelph forms occur in coral reefs at the base of the Eramosa beds and the true Guelph fauna comes in just above them. The transitional nature of the Eramosa beds is clear, and so Guelph species may be found at different horizons, but the Guelph formation in Ontario is clearly defined as a stratigraphic unit and is continuous with the New York formation above the Eramosa beds which are now recognized east of Niagara river.

The propriety of extending the use of the name Lockport over Ontario is supported by the continuity of the formation, the decided gradation where changes occur, and the marked faunal elements common throughout the formation. Of the one hundred and thirty-five species listed by the author for the Niagara dolomite of Ontario, twenty-five occur in the Lockport of New York state, and forty-eight occur in the Lockport at Thorold, only 8 miles from Niagara Falls. The northern coral development exemplified by occurrences at Cabot head and Fossil hill is quite different from the fauna of the more argillaceous dolomite at Lockport, New York, but important species are in common and it is well recognized that corals develop under special conditions and with special associates. The relationship of this coral fauna to the Louisville fauna is well marked, but the fact that the Guelph formation overlies the coral horizons of Bruce peninsula strengthens the conclusion arrived at by Butts[4] that the Louisville of Kentucky is upper Lockport in age (see also under Guelph, page 73).

The Lockport dolomite, or "Niagara limestone" as it was formerly called, forms the upper and most important member of the Niagara formation. In Ontario, this dolomite gives rise to the most conspicuous of the Palaeozoic cliffs, and outcrops in almost unbroken extent from Niagara falls (where the water pours over its very top) to Cockburn island at the head of lake Huron. The longest gap in the exposures is between Halfway

---

[1] For illustrations of the Lockport dolomite see Plates XXVIII A, XXIX A, XXX A, XXXII B, and XXXIII B.
[2] Kindle, E. M., Folio 190, p. 7.
[3] New York State Mus., Mem. 5, P. 14.
[4] Ibid., p. 99.

rock at the north end of Bruce peninsula and Fitzwilliam island. The only exposures of Lockport for the intervening distance of 22 miles are a few minor ones on Bear Rump, Flowerpot, Lucas, and Yeo islands, the re mainder of these islands and the other intervening islands being composed of Guelph dolomite.

The highest Lockport cliffs are those on Manitoulin island. Near Fossil hill, about 4 miles southwest of Manitowaning, and also at Campbell bay, the Lockport arises as cliffs 200 feet in height.

The Lockport is nearly a true dolomite in composition with the excep tion of the DeCew, Gasport, and Eramosa divisions (see chemical analyses) When fresh, the rock is generally of a grey, or a pleasing blue-grey colour but the familiar weathered surfaces are nearly white.

The total thickness of the Lockport dolomite at Niagara river is little less than 80 feet, or about 77 feet by Grabau's measurement. At Dundas, where the complete section is exposed, the thickness is about 135 feet. At Cabot head, the section is 160 feet thick and the top has not been recognized. On Manitoulin island, near Fossil hill, and also at Campbell bay as mentioned above, 200 feet of Lockport dolomite is exposed in section without any Guelph beds having been found. The above figures indicate the increasing thickness of the Lockport dolomite to the northwest, the lowest beds being probably older than the lowest Lockport beds in New York state. This is to be expected on the theory that the Lockport sea was an invasion from the north.

In the Niagara peninsula, the Lockport rests upon the Rochester shale. At Kelso the basal contact is with the Reynales dolomite and northward it is with the Cabot Head shales of the Cataract formation as described already. In a formation of such thickness, it is natural to expect that there will be considerable variation in bedding, character of the rock, fossil content, etc. The Lockport has such variations, but few divisions have more than local significance. The subdivisions and more detailed characters of the Lockport formation are described below. For the fauna, see table pages 64-68, and Plates XIV to XXI.

DeCew Waterlime Member.

The basal member of the Lockport dolomite in the Niagara peninsula consists of a bed of argillaceous, dolomitic limestone, 8 to 9 feet thick which was formerly extensively worked at Thorold for the manufacture of " natural rock cement " (for analyses, see page 112). The "waterlime" is fine grained and of a dark grey colour. As may be seen from Figure (page 22) this bed thins out at Stoney Creek and Hamilton to about 3 feet in thickness and has not been recognized to the west or north. At Kelso, the 3-foot bed immediately overlying the Reynales dolomite is typical Lockport dolomite in character and chemical properties (see page 173). At Limehouse, however, 6½ feet of thin-bedded, argillaceous, dolo mitic limestone was formerly worked for " natural rock cement." In these beds *Dictyonema retiforme*, *Atrypa reticularis*, and *Dalmania limulurus* occur. The first and last of these species are indicative of Rochester affinities and suggest that the lower beds at Limehouse are the equivalent of the DeCew beds. Other fossils in the DeCew are at Thorold

Sum Rept., Geol. Surv., Can., 1913, p. 186.

Strophonella patenta, Schuchertella subplana, ttbd *Camarotoechia (Steg r hynchus) neglecta;* and at DeCew falls the Rochester species, *Caryocrinites venatus.* Not only in fauna, but in composition also, the DeCew beds are closely related to the underlying Rochester shale. The waterlime is in fact a commingling of the dolomitic Lockport sediments with the argillaceous Rochester sediments. The characteristic nature of the DeCew beds may be seen in most of the exposures, and especially well in the Niagara gorge, along the New York Central and Hudson River railway north of Niagara university (see Plate XXX). Here and elsewhere the DeCew beds rest upon uneven eroded surfaces of Rochester shale, have churned structure, and present an eroded surface at the top into which the overlying dolomite fits. In general, the greater the irregularities at the top and bottom of the waterlime beds, the greater are the contortions within the beds themselves. Such characters point most clearly to a disturbance, which was capable of affecting what are now the 8 or 9 feet of waterlime beds from top to bottom, before the solidification of the original shales, and either just before or at the time of marine calcareous sedimentation. The probable conditions existing at this time will be discussed below.

Gasport Dolomite Member.

The term Gasport has been used by Kindle[] for a 9-foot bed of pure, crinoidal, semicrystalline limestone occurring above the DeCew beds at Gasport and Lockport, New York. According to Kindle "Westward from Lockport the limestone becomes gradually more magnesian", and "The section on the Canadian side at Niagara Falls shows 7 feet of hard, grey, subcrystalline, crinoidal limestone, sharply differentiated from the saccharoidal dolomite above and the 9 feet of argillaceous limestone below, and clearly the equivalent of the Gasport limestone member at Lockport, although it is dolomitic throughout".

" On the American side of the gorge the bed is well exposed along the footpath to the river just south of the railroad bridges, where it has its usual crinoidal character but is 20 feet or more thick and is wholly dolomitic. Both there and farther south along the Niagara Gorge railway it contains large lenticular masses . . . . They are lighter in colour than the enclosing rock and on weathering crumble to fine, buff-coloured powder. They are of irregular outline and differ considerably in size, the larger ones having a diameter of 18 to 20 feet and a height of 8 to 10 feet. It is believed that they represent reefs of bryozoans".

As will be seen from the above description crinoidal, semicrystalline characters are the distinguishing features of the Gasport dolomite on the Ontario side of Niagara river. It is to be expected that a member which is differentiated only by the presence of crinoidal characters will vary much in thickness from place to place. Thus, westward from the Niagara river crinoidal beds assigned to the Gasport are from 5 to 28 feet thick, as follows: Ontario side Niagara river, 14 feet (Kindle gives 7 foot); DeCew falls, not recognized, but at St. Catharines Power Company nearby 23 feet; Grimsby, 12 to 14 feet; Stoney Creek, 7 feet; Hamilton, about 5 feet; Ancaster, 15 feet; Dundas, 13 feet; Waterdown, 8 foot. Farther north the term Gasport does not seem applicable as crinoid columns

U. S. Geol. Surv., Folio 190, p. 7.

57237—5½

occur throughout the 70 feet of dolomite at Kelso, and elsewhere crinoic zones occur at different horizons in the Lockport.

The Gasport is 20 to 28 feet thick at Thorold [that] is composed large of coral reefs which are full of fossils. The varied life of these reefs suggested by the following numbers of genera and species which ha been recognized: corals, 9 genera, 13 species; crinoids, 2 genera, 2 specie brachiopods, 14 genera, 22 species; gastropods, 1 genus, 1 speci cephalopods, 1 genus, 1 species; trilobites, 2 genera, 2 species.

Undivided Lockport Dolomite.

The great mass of Lockport dolomite has not been subdivided exce into local facies and more or less local faunal zones. Leaving out t DeCew and Gasport members, already described, and the Eramosa bed described below, there is left the great central part of the Lockport dolomi of Niagara peninsula, all but the upper beds from Kelso to Fitzwillia island, and the whole formation on Manitoulin and Cockburn islands, be described in a general way.

The Lockport, as already stated, is dominantly a light grey, or bl semicrystalline, to fine-grained and compact, magnesian limestone appr[o] imating a true dolomite in chemical composition. Cavities and p[o] spaces are general. Chert nodules are plentiful at some local horizon and argillaceous beds occur fairly commonly in the Niagara peninsu[la] The dolomites are generally in beds 2 to 3 feet thick, but thinner beds [are] also much thicker beds are common. Jointing is general, is dominant vertical, and very irregular. Where the rock is exposed at the surfa[ce] erosion and solution by surface water has enlarged these joints to crac that are in some cases a foot or more across.

Beds Containing Chert.

Chert nodules, 2 or more inches across, occur at different horizons the Lockport formation. A zone of chert nodules is reported by Clarke a Ruedemann as occurring at the top of the thin Eramosa beds (determin by the writer) at Niagara falls. It is this chert horizon which is suppo to represent the location of the upper Shelby fauna (Guelph). Ch beds 5 feet thick, occur 10 feet lower in the section exposed near the Onta end of the international bridge.

The chert horizon has probably been eroded away in the section along Niagara escarpment between Niagara falls and Stoney Creek. So chert occurs at Stoney Creek in the beds immediately above the Gasp dolomite. At Mount Albion and Hamilton numerous chert nodules oc in the 60 feet of dolomite overlying the Gasport. From these nod[u] Colonel C. C. Grant collected many sponges. At Ancastor the chert b are 16 feet or more thick and at Dundas they are 40 feet thick; at b places they rest upon the Gasport dolomite.

No chert beds were noted in the sections between Dundas and Ca head. In the bluffs at Cabot head, a limited zone of chert nodules occ about 15 feet above the base of the Lockport formation and immediat above beds full of *Pentamerus oblongus*. Scattered chert nodules [are]

[1]New York State Mus, Mem. 5, p 14.

occur high up in the coral horizon at Cabot head and also on Manitoulin island. At Campbell bay, on Bayfield sound, 12 feet of fossiliferous chert beds occur about 180 feet above the base of the Lockport dolomite. A very remarkable chert horizon belonging near the top of the Lockport dolomite is exposed north of Old Quarry point, near the western end of Manitoulin island. Here, angular pieces of white chert 4 to 6 inches across, derived from the underlying beds, are scattered about all the surface of the rock. This chert zone is probably 15 feet thick.

Beds Containing Pentamerus Oblongus.

Pentamerus Oblongus, which is fairly common in the Clinton, Reynales beds of Niagara peninsula, appears to be nearly or quite absent for many miles to the north. At Owen Sound, however, the lower 5 feet of Lockport beds are crowded full of *Pentamerus oblongus*, the typical Lockport corals occurring just above. About 15 feet of basal dolomite at Wiarton is similarly crowded with this interesting brachiopod. Similar beds 4 or 5 feet thick occur at Lionhead and Dyer bay, and 2 feet of these beds occur at Cabot head. At Lionhead and Cabot head, however, 8 to 10 feet of thin, even-bedded dolomite occurs beneath the Pentamerus beds. The contact is sharp, the thin beds containing few fossils, and the Pentamerus beds being very fossiliferous. At Cabot head, the thin beds contain *Favosites favosus*, *Camarotoechia neglecta*, and a *Calocaulus* resembling *longispira*. It seems fairly certain, therefore, that the thin beds are basal Lockport. At Dyer bay, the basal beds are mostly obscured, and to the northward they are generally obscured as far west as West bay on Manitoulin island. Here, above 6 feet of soft, green Cataract shale, 20 feet of fine-grained, grey dolomite occurs with shale partings near the base. No fossils were found in these beds but they are probably basal Lockport.

Fossil Hill Coral Horizon.

The more northern occurrences of the Lockport dolomite are remarkable for the variety and numbers of their corals. These fossils are limited mostly to definite zones, which record suitable environment, as might be expected from a knowledge of the restricted conditions of temperature and pressure under which corals live. The corals at Fossil hill, about 7 miles southwest of Manitowaning, occur in great profusion about 70 feet up in the Lockport dolomite and appear to represent the acme of coral environment. It might not be expected that such favourable conditions prevailed widely at the geological moment recorded at Fossil hill, and yet similar coral assemblages are met with at approximately the same horizon at Cabot head, Campbell bay, Cockburn island, and elsewhere.

The " Barton Beds ".

The 80 to 90 feet of Lockport, overlying the chert beds and underlying the Guelph dolomite in the vicinity of Hamilton, are decidedly argillaceous and contain shale beds at some horizons. Spencer and Grant called these beds the " Barton beds," after the name of the township in which Hamilton is situated. The name " Barton " has never been

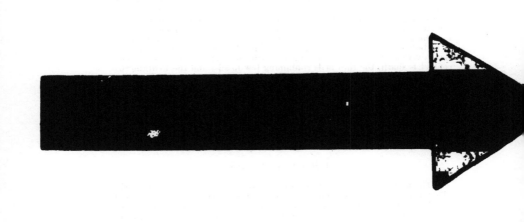

MICROCOPY RESOLUTION TEST CHART

(ANSI and ISO TEST CHART No. 2)

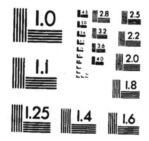

APPLIED IMAGE Inc

1653 East Main Street
Rochester, New York 14609 USA
(716) 482 - 0300 - Phone
(716) 288 - 5989 - Fax

recognized to any extent in geological literature as it was pre-occupied in the Tertiary of England; it has, however, considerable local significance. As will be seen later, the upper 35 feet of the " Barton beds " belong to the Eramosa dolomite. The lower part is well-defined by its argillaceous character from Mount Albion to Ancaster.

Lockport Dolomite not Differentiated.

Northward from the Niagara peninsula, the greater part of the Lockport dolomite is undifferentiated. It differs considerably in bedding and characters, but is generally thick-bedded to massive, typical dolomite, blue, grey, or less commonly buff on fresh fracture, and weathering white.

[1]Eramosa Member.

The Eramosa beds consist characteristically of thin, even-bedded, argillaceous, arenaceous or bituminous dolomites. They are sparingly fossiliferous. These beds occur at the top of the Lockport formation throughout the western peninsula of Ontario and on the islands to the north as far as Fitzwilliam island. Good sections of the Eramosa beds, however, are few, the following being the best known (Figure 3): Niagara Falls, New York, 12 feet of thin, even-bedded dolomites with bituminous partings, their top being at the crest of the American falls; Ancaster, not well-differentiated but about 35 feet of brown-grey dolomite, with bedding planes marked by change of colour; Dundas, about 35 feet of dark blue, argillaceous dolomite, shaly at base and with bituminous partings at top; Guelph, 30 feet or more of a chocolate brown, bituminous shale and dolomite; Wiarton, about 50 feet of dark grey, thin-bedded dolomite, very cleavable and even-bedded in the upper 15 feet, which are quarried for flags and dimension stone; Halfway rock, 30 feet of thin-bedded dolomite. The upper part of the Eramosa beds are seen in the lower part of the " flowerpots " of Flowerpot island (see Plate I). On the western end of Fitzwilliam island, typical Eramosa beds occur at the top of the Lockport; their thickness is probably about 30 feet. On Manitoulin island the Eramosa beds have not been recognized and appear to have been eroded away.

The Eramosa beds are marked by dome structure, the dimensions of the domes varying from a few feet to 100 or 200 feet across, with a centre elevation of 15 to 20 feet. Good examples of the larger domes occur at Guelph, where one exposure south of the prison farm has the top of the dome stripped away, revealing a dome-shaped coral reef which in this case at least was the cause of the structure in the Eramosa beds.[2] This reef contains a fauna which is transitional from the Lockport into the Guelph.

Few fossils have been found in the Eramosa beds, excepting at Guelph[3]. The revised list of those collected by the writer from an horizon about 25 feet below the top of the Eramosa beds is as follows: *Monomorella* cf *orbicularis* Billings, *Orbiculoidea subplana* (Hall), *Orthis?* near *tenuidens*

[1]Geol. Surv., Can., Mus. Bull. 20, p. 1.
[2]Ibid. p. 2.
[3]Ibid. p. 3.

Hall, *Camarotœchia (Stegerhynchus) whitei* (Hall)?,*Spirifer radiatus* Sowerby? *Whitfieldella nitida* Hall?, *Orthoceras, Eusarcus logani* Willia? s.

The approximate chemical co? position of the fossil horizon at Guelph as determined by E. Poitevin of this Survey, is as follows: insoluble ? atter including iron oxide, 50 per cent; bitu? inous ? atter, 25 per cent; calciu? carbonate with a trace of ? agnesia, 25 per cent.

As already stated, the lower fossil horizon, containing *Monomorella* cf *orbicularis*, is suggestive of the lower Shelby horizon of New York state, and the true Guelph above is probably represented in New York by the upper Shelby horizon which, therefore, would ? ark the base of the Guelph for? ation.

Distribution of Lockport Fauna.

Genera and species.	Western New York.	Niagara river.	Thorold.	De'ew fluis. {Jasport, Del'ew,	Higher beds, {Grimsby, Gasport boulders,	Stony C'reek.	Hamilton. {Ginsport, C'hert, Higher,	Dundas.	Limehouse, Lower 15 feet,	Rockwood, Upper,	Guelph, Heel near top and Eramosa,	Rolaporo, 50-100 feet up.	Owen Sound, Lower, 30 feet up.	Wiarton, Lower 8 feet, 125 feet up.	Purple valley, Upper,	Barrow bay, Lower,	Stokes bay, Upper,	Island, {hot head, 30 feet up. High up.	Fitzwilliam Island, {Near base, Higher, 170 feet up.	Hungerford point.	Tamarack cove, Middle,	South bay, Upper,	Fossil hill, 80 ft. up to 100 ft. up.	Sundfield, Middle.	Michael bay, Upper,	Providence bay, Upper, 90 ft. up.	Kingawong lake,	Quarry point, Upper,	Lockburn Island, 100' ft. up.
Plants.																													
Buthotrephis gracilis (Hall)																													
Buthotrephis gracilis crassa Hall																													
Porifera.																													
...gia praemorsa (Goldfuss).																													
...pha.																													
Dictyonema retiforme (Hall)																													
...ip ma gracile Hall.																													
Actinostroma whi ...aesi ...gee Parks																													
Clathrodictyon vesiculosum Nicholson and Murie.																													
C... ...ipn variolare (Von Rosen).																													
Clathrodictyon striatellum (D'Orbigny).																													
Clathrodictyon ostiolatum ...lson).																													
Lophiostroma magnum Parks.																													
Rosenella? manitoulinensis Parks.																													
Stromatopora constellata Hall.																													
Stromatopora antiqua (Nicholson and Murie)																													
Ceramoporella? irregularis (Whitfield).																													
Anthozoa.																													
Enterolasma caliculum (Hall)																													
Lindstroma spongaxis (Rominger)																													

Zaphrentis stokesi Edwards and Haime
Amplexus shumardi (Edwards and Haime)
Pycnostylus guelphensis Whiteaves
Pycnostylus elegans Whiteaves
Ptychophyllum stokesi Edwards and Haime
Cystiphyllum niagarense (Hall)
Chonophyllum canadense (Billings)
Strombodes diffluens Edwards and Haime
Strombodes eximius Billings
Strombodes pentagonus Goldfuss
Cyathophyllum thoroldense Lambe
(Synaptophyllum multicaule (Hall)
Diplophyllum caespitosum Hall
Cystiphorolites major (Rominger)
Syringopora bifurcata Lonsdale
Syringopora retiformis Billings
Syringopora verticillata Goldfuss
Syringopora fibrata Rominger
Cannapora junciformis Hall
Vermipora niagarensis Rominger
Favosites constrictus (Hall)
Favosites favosus (Goldfuss)
Favosites hisingeri Edwards and Haime
Favosites hispidus Rominger
Favosites niagarensis (Hall)
Paleofavosites asper D'Orbigny
Syringolites huronensis (Hinde)
Thecia major Rominger
Alveolites thoroldensis Parks
Alveolites labeelei Edwards and Haime
Cladopora fibrosa Hall
Cladopora laqueata Rominger
Cladopora reticulata Hall
Coenites crassus (Rominger)
Coenites juniperinus Eichwald
Coenites laminatus (Hall)
Coenites ramulosus (Hall)
Striatopora flexuosa Hall
Halysites catenularia (Linnaeus)
Halysites catenularia microporus (Whitfield)
Lyellia americana Edwards and Haime
Lyellia affinis (Billings)
Lyellia superba (Billings)
Lyellia decipiens Rominger
Heliolites elegans Hall
Heliolites interstinctus (Linnaeus)
Heliolites megastoma (McCoy)

[1] Classification revised by Chadwick.

Distribution of Lockport Fauna—(Continued).

Genera and species.	100 ft. up. Lockburn island.	Upper, Quarry point.	80 ft. up. Kagawong lake.	Upper, Providence bay.	Upper, Michael bay.	Middle, Sandfield.	80 ft. up to 100 ft. e. c. Fossil hill.	Upper, South bay.	Middle, Timmins's r. e.	Hungerford point.	Higher, 170 feet up. / Near base. (Fitzwilliam island.)	High up. 80 feet up. Basal. (Cabot head.)	Upper, Stokes bay.	Lower, Barrow bay.	Upper, Purple valley.	125 feet up. / Lower 8 feet up. (Wiarton.)	Lower. 80 feet up. (Owen Sound.)	50–100 feet up. Kolapore.	Reef near top and Fernton. Guelph.	Upper, Rockwood.	Lower 15 feet. Limehouse.	Dundas.	Higher, Chert. (Hamilton.)	Gasport.	Stony creek.	Higher beds, Gasport boulders. (Grimsby.)	De New falls. Gasport.	Thorold.	Niagara river.	Western New York.
Anthozoa—Con.																														
Heliolites pyriformis Guettard																														
Heliolites spiniporus Hall.																														
Millites subtubulatus (McCoy)																														
Plasmopora follis Edwards and Haime																														
Protarea walkeri (Spencer)																														
Cystoidea.																														
Caryocrinites ornatus Say																														
do.																														
Crinoidea.																														
Periechocrinus speciosus (Hall)																														
Ichthyocrinus laevis (...																														
Eucalyptocrinus caelatus laevis Grabau and Shimer																														
Dendrocrinus reductus (Hall)																														
Bryozoa.																														
Labechia euphrentica Hall.																														
Brachiopoda.																														
Orbiculoidea subplana (Hall).																														
Schizotreta tenuilamellata (Hall)																														
Dictyonella coralliafera (Hall).																														
Strophaeodonta profunda (Hall).																														
Strophonella striata (Hall)																														

Platystrophia biforata (Schlotheim)
Bilobites acutifolus (Ringueberg)
Dalmanella elegantula (Dalman)
Rhipidomella hybrida (Sowerby)
Conchidium occidentale Hall
Stricklandinia manitoupensis Williams n. sp.
Pentamerus oblongus Sowerby
Clorinda ventricosa (Hall)
Rhynchotreta cuneata americana (Hall)
Camarotrechia neglecta (Hall)
Camarotrechia whitei (Hall)
Camarotrechia ? acinus (Hall)
Camarotrechia ? indianensis (Hall)
Atrypa reticularis (Linnaeus)
Atrypa nodostrata Hall
Cyrtia meta (Hall)
Spirifer niagarensis (Conrad)
Spirifer radiatus (Sowerby)
Spirifer crispus (Hisinger)
Spirifer eudora (Hall)
Homoeospira aprimformis Hall
Whitfieldella nitida (Hall)
Whitfieldella nitida oblata (Hall)
Whitfieldella intermedia (Hall)
Whitfieldella cylindrica (Hall)

Pelecypoda.
Matheria recta (Hall)
Amphicoelia leidyi Hall
Mytilarca acutirostra (Hall)
Streptomytilus aphaea (Hall)
Goniophora speciosa Hall

Gastropoda.
Eotomaria kayseri Clarke and Ruedemann
Coelocaulus longispira (Hall)
Coelocaulus bivittatus (Hall)
Holopea guelphensis Billings
Diaphorostoma hemisphericum Hall
Diaphorostoma niagarense (Hall)

Conularida.
Conularia niagarense Hall

Cephalopoda.
Orthoceras imbricatum (Wahlenberg)
Orthoceras bruxensis Williams n. sp

¹ Classification revised by Chadwick.

Destribution of Lockport Fauna—Concluded.

Genera and species.

Localities (column headers):
Western New York; Niagara river; Thorold; Del'ew falls (Grasport); Higher beds (Grasport boulders) Grimsby; Stonoy Creek; Gasport, ('hert,) Higher, Hamilton; Dundas; Lower 15 feet, Limehouse; Upper, Rockwood; Reef near top and Firmount, Guelph; 50-100 feet up, Kolapore; Lower, 50 feet up, Owen Sound; Lower 8 feet, 125 feet up, Wiarton; Upper, Purple valley; Lower, Harrow bay; Upper, Stokes bay; Basal, 30 feet up, High up, Cabot head; Near base, Higher, 120 feet up, Fitzwilliam island; Hungerford point; Middle, Tamarack cove; Upper, Mouth bay; 80 ft. up to 100 ft. up, Fossil hill; Middle, Mandfield; Upper, Michael bay; Upper, Providence bay; 80 ft. up, Kuawwong lake; Upper, (?) point; 100? ft. up, Khuru Island.

Cephalopoda—Con.
Dawsonoceras annulatum (Sowerby).
Kionoceras cancellatum (Hall).
Protokionoceras trusitum (Clarke and Ruedemann).....
Trochoceras costatum (Hall)......
Discosorus gracilis Foord......
Huronia vertebralis Stokes.
Trilobita.
Bumastis ioxus (Hall).....
Goldius niagarensis (Hall).....
Calymene niagarensis Hall.....
Sphaerexochus romingeri Hall.....
Dalmanites limulurus (Green).....
Eurypterida.
Eusarcus logani Williams.

Note.—r, reported by Parks.   q, reported by Grabau.   q, found by author.   L, reported by Lambe.   s, reported in Bassler's Index.
¹Classification revised by Chadwick.

CORRELATION OF THE NIAGARA GROUP.

The Rochester shale is an extension westward of the Rochester shale of New York state and is generally correlated with the Osgood limestone, and shale of west Tennessee, Ohio, and Kentucky and, according to Bassler, with the West Union limestone of southern Ohio and eastern Kentucky and the upper St. Clair limestone of northern Arkansas. Twenhofel[1] correlated his Chicotte formation of Anticosti island "on stratigraphic grounds..........with the Irondequoit-Rochester of the New York section." The Rochester has been correlated by Schuchert and Twenhofel with the McAdam formation of Arisaig, which McLearn[2] correlates with "the Salopian" of Great Britain "without regard.........to its component divisions".

The Lockport formation of Ontario and New York state is to be correlated with the Laurel dolomites, the Waldron shale, and the Louisville magnesian limestone of Kentucky[3], the Glenkirk limestone of west Tennessee, the Hopkinton dolomite of Iowa and Minnesota, and a part of the Bainbridge limestone of east Missouri and Illinois[4]. Savage[5] has recently claimed that the Mayville limestone of Wisconsin is the equivalent of some part of the Alexandrian series of Illinois and Missouri and the author correlates the Racine of Wisconsin with the lower Guelph of Ontario; thus only the Waukesha beds, including the Byron, lower and upper coral beds, remain to be correlated with the Lockport dolomite of New York and Ontario. These subdivisions may yet be traced into Ontario, the chert horizon containing fossils at Campbell bay, Manitoulin island, suggesting strongly the upper Coral beds[6].

The long recognized Niagara strata of lake Timiskaming region have been shown by the author to contain a coral horizon closely resembling that at Fossil hill, thus indicating its relationship to the Lockport. Schuchert[7] has correlated with the Louisville, deposits occurring along the west shore of James and Hudson bays, in the lake region of Manitoba, in southeastern Alaska, and on nearly all the islands of the Arctic archipelago north of Canada. Although much remains to be learned of these little studied regions, the fossils in many cases suggest Lockport affinities. For example, coral reefs containing "*Favosites gothlandicus* Lamarck, *Heliolites interstincta* (Linn.), and *Halysites catenulatus* (Linn.) var.,"[8] found on the 141st meridian between Yukon and Alaska by the author, while assistant to D. D. Cairnes, suggest the Lockport coral horizon. The fossils collected by A. P. Low[9] in 1904, on Southampton island, Hudson strait, include "*Favosites gothlandicus, Syringopora verticillata, Halysites catenularia,* and *Plasmopora follis*," all of which are typical Lockport corals.

The Lockport is to be correlated with the Moydart[10] formation of Arisaig and the upper Wenlock of Norway.

[1] Ibid, p. 21.
[2] Am. Jour. Sc., vol. XLV, 1918, p. 135.
[3] Kentucky Geol. Surv., ser. IV, vol. III, pt. II, 1914-15, pp. 84, 46, 99.
[4] U.S. Nat. Mus. Bull. No. 92, vol. II, 1915, pls. 3 and 4.
[5] Bull. Geol. Soc. Am., vol. XXVII, No. 2, 1916, p. 305.
[6] Geol. of Wis., vol. II, p. 351.
[7] Bull. Geol. Soc. Am., vol. XX, 1909, pl. 67.
[8] Cairnes, D. D., Geol. Surv., Can., Mem. 67, p. 74.
[9] Cruise of the Neptune, 1906, Appendix IV, pp. 325-6.
[10] Geol. Surv., Can. Mem. 60, 1914, p. 80.

NIAGARA SEDIMENTATION.

From the foregoing evidence, it is clear that Niagara time commenced with seas which in places received large quantities of mud washed in from the surrounding land, the deposits forming the Rochester shale. Either the younger formations to the north were elevated above the land and were actively eroded during Rochester sedimentation or else (as believed by Chadwick) an emergence followed Rochester deposition during which not only Rochester sediments but younger sediments to the north were eroded away. With the succeeding submergence, came an overlapping of the dolomitic deposits of the Lockport sea from the north extending over higher and higher Rochester deposits (according to Chadwick) well into New York state. On the basis of this overlap, the DeCew beds must be regarded as reworked Rochester shale, their final condition being probably due to the action of currents and waves of the encroaching sea, upon loose materials derived from land erosion. The fossils contained in the DeCew beds would thus be, either dead forms contained in the Rochester shales or forms introduced under the new conditions of Lockport sedimentation. As a matter of fact their affinities are mostly with the Rochester and they do not indicate any apparent wear.

Kindle[1] has shown that beds similar to the DeCew waterlime are being formed to-day in the bay of Fundy as the result of deposition and scour, and although the widespread DeCew beds may not be compared directly with the modern formations of the bay of Fundy, the origin of the latter beds is instructive. The higher dolomites were deposited in wide-spreading seas, which were of such depth and temperature as to promote the growth of coral reefs even within the Arctic regions. A partial emergence of the land caused an inflow of sufficient detrital material to give the Eramosa beds their argillaceous and arenaceous character, these beds being deposited against and over coral reefs whose development they arrested. This period of diminishing seas closed Niagara time, the succeeding deposits being Guelph.

GUELPH FORMATION.[2]

DESCRIPTION, EXTENT, AND SECTIONS.

The Guelph is a great dolomite formation of fairly uniform character, which overlies the Lockport dolomite and is overlain by the Salina formation. The weathered rock is light grey or cream-coloured and generally has a porous saccharoidal texture. Some beds, however, are compact, fine-grained, and very hard. The beds vary from a few inches to several feet in thickness, the thicker beds being most common.

The Guelph formation underlies a belt of country in Ontario extending from Niagara river to the head of Bruce peninsula and occurs on the islands to the north as far as Fitzwilliam island. In Niagara peninsula this belt averages about 5 miles across, and from Hamilton to Chiefs point the average is about 13 miles. On Bruce peninsula the Guelph is divided into areas, isolated by denudation. Unlike the underlying Lockport, the Guelph dolomite does not occur as prominent cliffs except at the northern end of

[1] Bull. Geol. Soc. Am., vol. 28, 1917, pp. 323-334.
[2] For illustrations of the Guelph dolomite, see Plates I, XXVIIA, XXXIIA, XXXIII, XXXIVB.

Bruce peninsula and in the islands immediately to the north. Elsewhere the rock is exposed as shelving ledges along the shore of the west side of Bruce peninsula, as glaciated exposures protruding through the surface deposits over large areas of Bruce peninsula and to a less extent in Grey and Wentworth counties, and in the valleys of streams, notably the Rocky Saugeen and Grand rivers and the upper waters of Twentymile creek (see Maps 1714 and 1715 for exposures).

Owing to the lack of complete sections, the thickness of the Guelph formation is difficult to measure. Logan estimated it on the basis of dip to be about 160 feet thick in the vicinity of Breslau.[1] He moreover, described the section at Elora as 82 feet thick and the section at Fergus as 20 feet thick "which would underlie the preceding." He placed the strata farther up the Grand river as still lower.

The writer followed the section from Fergus to Elora along the Grand river at low water and found that the river descends at a higher angle than the dip of the formation and in consequence the top beds at Fergus are about 10 feet higher than the top of the section at Elora and about the same as the top beds at Salem 1 mile northwest of Elora. The exposures above Fergus appear similar to those at Fergus and give little additional information.

The Salem-Elora section represents all the strata recognized and was found to measure about 82 feet as stated by Logan. The section consists for the most part of compact, fine-grained dolomite and is as follows in ascending order: 13 feet of thin-bedded dolomite containing *Favosites favosus*, *Pycnostylus guelphensis*, and *stromatoporoids*; 19 feet of beds 2 to 3 feet thick containing *Megalomus canadensis* in the lower 12 feet; 8 feet of beds 3 to 4 feet thick containing *M. canadensis*; 20 feet of hard, fine-grained, light buff beds about 2 feet thick; 12 feet of thinner beds at the top of the section containing *Conchidium occidentale*, *Colocaulus* sp., and *M. canadensis*. In places the thicker beds appear quite massive and no division extends very far laterally. Ten feet of higher strata occur at the top of the section at the steel bridge over a branch stream near Salem and consist of fine-grained, compact, rather thick beds containing *M. canadensis* about 8 feet below the top.

The best method now available for measuring the thickness of the Guelph is by reference to well records. Measurements cannot be derived directly, however, as it is difficult to recognize in well records the division between the Lockport and Guelph formations. The Lockport has, however, been measured in section both at Niagara river (77 feet) and Dundas (155 feet) (see page 58) and in a well sunk at Guelph[2] the "blue slate" (= Fermosa beds) is 50 feet and the dolomite below is 110 feet thick. The Lockport is about 150 feet thick. Assuming these measurements to hold approximately within a few miles radius of the locality where they were made, we arrive at the following thicknesses for the Guelph: on lot 2, concession IV, Willoughby township, Welland county,[3] the combined thickness of the Guelph and Niagara dolomite is 220 feet; taking the Lockport as 80, the Guelph is 140 feet thick. As this well is only 6 miles from Niagara the measurement may be considered very nearly correct. A well at Kitchener[4] gives what may be interpreted as the

[1] Geology of C....
[2] Rept. Ont. Bur...... nes, 1903, p. 150.
[3] Well record, Jour..... Min. Inst., vol. III, p. 77.
[4] Geol. Surv., Can.... t., pt. II, p 42A.

thickness of the Guelph and Niagara dolomite as 240 feet. Taking the thickness of the Lockport as 150 feet as at Guelph, the Guelph formation would be 90 feet thick. Considerable difference of thickness is thus indicated and it is probable that this is to be explained by an erosion surface at the top of the Guelph, which is clearly suggested by the varying thicknesses of the combined Guelph and Lockport dolomites seen in the cross section from Dunnville to Grimsby (see Map 171E) and in the well records of Kent county (see Figure 4). It will be noted that 90 feet compares rather closely with the section as measured at Elora. The Elora section is certainly not complete, but may not lack as much as Logan supposed. Northward, little reliable information is at hand as to the thickness of the Guelph formation

The base of the Guelph dolomite is clearly exposed in the eastern part of the city of Guelph, where it rests on the Eramosa beds. The lowest bed is about 2 feet thick and is bituminous, with a smell resembling that of petroleum. Thinner, light brown, nodular beds succeed, about 12 feet being exposed along the Canadian Pacific railway near Eramosa river. *Pycnostylus guelphensis* occurs about 8 feet above the base. The base of the Guelph is also exposed at Niagara river, at Dundas, in the vicinity of Wiarton, at Stokes bay, at Halfway rock (about midway between Cabot head and Tobermory), and on Bear Rump, Flowerpot Co., and Fitzwilliam islands.

Between the city of Guelph and Niagara river the weathered basal Guelph beds, averaging about 2 to 4 feet in thickness, are brown, have bituminous partings, and a smell resembling that of petroleum. They are also commonly full of cavities which contain at Niagara gorge "pearl spar or dolomite". In the vicinity of Wiarton and northward the basal Guelph is massive, full of cavities and pore spaces but does not have bituminous partings or an oily smell.

The Standard White Lime quarry in the southwestern part of the city of Guelph is in strata that are probably from 40 to 60 feet above the base of the Guelph formation. The section includes about 20 feet of creamy white, porous, saccharoidal dolomite, the beds being 3 to 4 feet thick near the base and thin (probably due to weathering) near the top. *Stromatoporoids* and *Calocaulus* (more than one species) are common near the base, *Pycnostylus guelphensis* occurs about 8 feet up, and *Trimerellas* and *Megalomus canadensis* occur near the top.

As already stated, beds high up in the Guelph formation are represented at the top of the Salem-Elora section. The sections at Galt and between Galt and Glen Morris (on the Grand river about 6 miles below Galt) include the highest Guelph beds known to outcrop in Ontario. The highest beds of all are exposed at or near water-level in the river above Glen Morris. The dolomite is heavy water-stained, occurs in beds about 10 inches in thickness, and is fine-grained and very hard. On fresh fracture it is buff grey. *Megalomus canadensis* was found in most of the exposures. Because of the isolated outcrops and the variations in the dip of the bedding the thickness of exposed strata could not be determined. Logan[2] has described the Galt section as consisting of "twenty feet of yellowish-white and greyish-white, crystalline, thick-bedded dolomite" at the base, succeeded upward by "thirteen feet of pale buff or white dolomite", and "eighteen

[1] Bull. N.Y. State Mus., No. 45, 1901, p. 112.
[2] Geol. Surv., Can., 1863, p. 339.

73

et of hard, thin-bedded, bluish dolomite" at the top. He continues,
"the whole mass holds fossils, but these, in immediate vicinity of Galt.
are most abundant in the twenty feet of p off, thin-bedded dolomite
in the middle of the section". The strata ownstream are evidently
included by Logan in the upper beds and probably do not represent more
than the 20 feet allowed by him, as the river falls nearly on the strike of the
formation. 31 feet of section is exposed at Galt above the river bed.

At Niagara falls, all the rock abo e the back of the falls is Guelph
According to the section given by Clarke and Ruede on 44 feet of
Guelph dolomite is represented in the vicinity of th alls, the ou rops
being a the car bath on the gorge route (basal), on Goat island, and
on T e Sisters islands. This section is s follows in ascending order:
"8 feet of brown, thin-bedded dolomites with rough surface and black
partings", 19 feet not exposed. "25 feet brown dolomites exposed at
Goat island and Three Sisters islands with summit near head of rapids.
Full of cavities with Stromatopora, Halysites. Weathers very scraggy.
Uppermost bed (4 feet) sandy."

Besides the Fergus-Elora section, a fair section is exposed on the
Rocky Saugeen river at Hayward falls just below Rocky Saugeen post-
office. The gorge in the rock is 30 to 40 feet deep and in all 90 feet of rock
is exposed. The sidefaces are badly weathered and caves occur near the
tail-race of a mill. The rock is thick-bedded to massive and contains
numerous fossils, most of which were found above the middle of the section.
Favosites hisingeri?, Pycnostylus guelphensis, Megalomus canadensis, Conchidium occidentale, and Stromatoporoids were recognized.

Good outcrops also occur at Durham, from which many fine specimens
were obtained by Mr. J. Townsend and other early collectors.

In the icinity of Wiarton, as already mentioned, the base of the
Guelph is ll exposed. The best exposures are about two miles northwest from Wiarton on the road to Oliphant where the Framosa beds
are quarried. The actual contact with the base of the Guelph is obscure,
5 or 6 feet of strata being talus-covered. About 15 feet of Guelph strata,
however, are exposed in a cliff to the north and consist of porous scraggy,
rather thick-bedded dolomite containing Pycnostylus guelphensis, Spirifer
eudora, and Stropheodonta sp. At a zinc prospect about 4 miles northwest
of Wiarton on the north half of lot 30, concession 11, Albemarle East
township, Bruce peninsula, an open-cut has been made in prospecting
for zinc blende. The cut starts in rock that is about 40 feet above the
base of the Guelph and is 30 foot deep, thus supplementing the section
just mentioned. The rock, which is light grey in colour, is very porous
and scraggy and full of cavities, many of which contain zinc blende. In
the dump from this excavation, many fossils have been obtained; these
are quite unlike Guelph assemblages found elsewhere in Ontario and
very similar to the fauna of the Racine beds of Wisconsin. The outstanding
features of the fauna are the large number and variety of cephalopods
present. Most of these cannot be definitely determined because of their
poor preservation, those recognized being given in the following table
(pages 77-79). An unidentified Phragmoceras is also very common. The
most typical Guelph species present is Pycnostylus guelphensis.

Higher Guelph strata occur along the western side of Bruce peninsula,
and many specimens of Megalomus canadensis occur in the strata along

Clarke and Ruedemann, p. 14.
57237—6

the shore between Stokes bay and cape Hurd. Beds at Stokes bay within the lowest 50 feet of the Guelph formation are crowded with very large crinoid columnals. Some of these are three-quarters of an inch in diameter. It was formerly thought that the Guelph formation in Ontario was almost free from crinoid remains, and this is mostly true for the southern and better known outcrops. In the north, however crinoid stems are common and some of the lower beds are almost entirely made up of them.

The fishing islands west of Oliphant and the islands north of Tobermory, have excellent Guelph exposures. The lower beds, as for example at Halfway rock, Bear Rump, Flowerpot, Middle, Lucas, Yeo, and Fitzwilliam islands, have been recognized primarily by their stratigraphic position above the Eramosa beds. These upper Lockport beds extend to the northeast, at the localities mentioned, from beneath the Guelph capping rock, with the exception that on Bear Rump island the Eramosa beds project to the southwest from beneath the higher strata. Guelph fossils were found at each of the above-named localities, however, and thus the age of the rock was verified. For higher beds, age determination depends upon location and fossil evidence. Casts of species of *Trimerella* and *Rhinobolus* are of great value in determining the age of the northern Guelph outcrops.

Reef structures occur on Middle or "Plucky island" and on Lucas and Yeo islands. In fact it is to massive reef structure that these small islands owe their preservation.

"Plucky", or middle island as it is known on the chart, is a mere islet situated about halfway between Tobermory and Flowerpot island and is not shown on the geological map. It appears as a mass of tumbled blocks of dolomite, 30 or 40 feet high, but on closer inspection it proves to have a core of solid rock from which the blocks have fallen, being undermined by wave action. The solid rock is a *stromatoporoid-coral* reef. The reef building corals were *Pycnostylus guelphensis* and *Pycnostylus elegans*. Along with the reef builders, *Favosites favosus*, *Rhinobolus guelphensis*, and *Trimerella grandis?* occur. The top of Lucas island appears to be the remains of a *stromatoporoid* reef resting on Eramosa beds. The maximun elevation is about 30 feet. Yeo island is low, probably not more than 20 feet high at any place, and is a domed-reef structure resting on more or less domed Eramosa beds which are incised by wave action on the northeast angle. The reef builders were *Stromatoporoids*, and these in some cases had their bases below the top of the Eramosa beds, the structure thus resembling the transition coral reef at Guelph described under Eramosa beds. Brachiopods and gastropods lived in association with the reef builders; *Atrypa reticularis*, *Whitfieldella nitida?*, and *Pleurotomaria?* being noted.

The most northerly occurrence of Guelph is on the southwestern extremity of Fitzwilliam island. The rock capping the ends of the numerous rock projections and the small islands just off-shore is Guelph. The typical, hummocky, Eramosa beds are exposed on the mainland at the base of the scraggy, thick-bedded Guelph and a very few typical Guelph fossils were found. Among these, *Rhinobolus galtensis* is the most reliable guide to the Guelph. No Guelph strata have been identified on Manitoulin island. Some of the rock at South Baymouth is much like the Guelph, but although the shore was studied both east and west.

no Eramosa beds were identified and no Guelph guide fossils were found. For Guelph fossils see table page 000, and Plates XXII to XXVII.

CORRELATION.

The Guelph formation extends into New York state, including, as already stated, the upper 43 feet of the section exposed along the Niagara river and according to Chadwick probably 100 feet of higher beds. The Eramosa beds have been recognized by Chadwick and the author in the barge canal at Rochester and the dolomite above is typically Guelph. This accords with the maps of New York state. As already stated under "Lockport formation," the upper Shelby and the overlying beds may be considered the New York equivalent of the Guelph formation of Ontario. The Racine beds of Wisconsin and Illinois contain, according to Chamberlain[1], a cephalopod fauna very similar to the fauna in the lower Guelph near Wiarton, and also many distinctly Guelph genera and species, such as *Megalomus canadensis, Dinobolus Conradi,* and *Trimerella cf grandis.* The author, while visiting the more important Wisconsin and Illinois occurrences of the Racine beds in 1914, was impressed with the decided Guelph characters of the Racine, and moreover, accompanied by Professor Stuart Weller of Chicago university, he was able to point out the great similarity of the lower beds exposed in a quarry at the locality south of Chicago known as "Stony Island," to the Eramosa beds of Ontario. The general thin-bedded character of 15 feet of these beds and the vugs of soft bitumen in some lower beds were very suggestive of the general horizon of the Eramosa beds, and the presence of a poor specimen of *Pycnostylus guelphensis* in the upper beds of the quarry was additional evidence that the contact between the Guelph and Niagara formations is exposed here. The general northward dip of the strata would readily allow for the Racine and higher beds occurring above this horizon at Chicago, and northward. More field work is necessary before the Guelph-Niagara problem is cleared up in Wisconsin and Illinois, but in the meantime the author prefers to correlate the Racine and the overlying "Guelph beds" of Chamberlain with the Guelph of Ontario.

Kindle[2] has reported Guelph fossils from limestone at Huntington, Indiana, which according to Foerste[3] overlies the Louisville limestone. Strata of Guelph age are doubtless present. Elsewhere in the United States, few occurrences of strata of undoubted Guelph age are reported. *Pycnostylus guelphensis* and *Pycnostylus elegans,* found by Dowling[4] on Ekwan river, west of James bay, probably indicate the presence of Guelph strata in this region. Loose glacial boulders containing fossils suggesting Guelph age were found by Hume north of lake Timiskaming. These probably came from James Bay region. *Pycnostylus guelphensis* was found by Tyrrell[5] on Davis point, lake Manitoba, suggesting uppermost Niagara or Guelph strata at this vicinity. It is altogether probable that with further detailed work, other occurrences of Guelph strata will be found in northern Canada. Twenhofel and Schuchert[6] have correlated the

[1]Geol. of Wis., vol. II, 1873-77, pp. 374-377.
[2]Ind. Dept., Geol. and Nat. Res , 28th Ann. Rept., 1904, p. 408.
[3]Ind. Geol. and Nat. Res., 28th Ann. Rept., 1904, pp. 34 and 35.
[4]Palæozoic fossils, vol. III, pt. IV, p. 247.
[5]Palæozoic fossils, vol. III, pt. IV, p. 284.
[6]Am. Jour. Sc., vol. XXVIII, 1909, p. 148.

Stonehouse for 1ation of Arisaig, No1a Scotia, with the Guelph, and both with the Ludlow of Europe.

GUELPH SEDIMENTATION.

It is generally believed that Guelph strata were deposited in a so 1 e-what restricted sea of increasing salinity. Thick-shelled for 1 s such as *Trimerella* and *Megalomus* are generally taken as being indicative of high salinity and the restricted areas of known Guelph outcrops suggest narrow sea-ways. Evidence is accu 1 ulating, however, in support of Grabau's theory that the Salina was derived largely fro 1 erosion of the Guelph and Niagara for 1 ations. By a careful study of sa 1 ples taken fro 1 wells east of lake St. Clair, it is seen that there is no indication of an erosion surface anywhere in the Guelph-Niagara dolo 1 ites, but that where these are unusually thick, the upper beds have distinctive characters, of colour and hardness. Although this may be due to lenticular deposition, it strongly suggests erosion in pre-Salina ti 1 e. Although la 1 ellibranchs, brachiopods, and gastropods are the 1 ost striking Guelph fossils, coral reefs are now known to be co 1 1 on, and great colonies of crinoids are common in so 1 e places. Cephalopods were extre 1 ely i 1 portant during the early stages of the Guelph sedi 1 entation in the region of Bruce peninsula and Wisconsin, and were fairly well represented throughout the period. The closing event of Guelph ti 1 e appears to have been an e 1 ergence and erosion period during which a part of the lower Great Lakes region may have re 1 ained a partly enclosed sea.

Distribution of the Guelph Fauna.

Genera and species.	West end Fitzwilliam island. Lower.	Middle island. Middle.	Cove and Echo islands. Upper.	Cape Hurd. High up.	Tobermory. Upper.	Halfway rock. 100 ft. up.	Stokes bay. Basal beds.	Fishing islands. Lower.	Zinc prospect. Near Wiarton (Hanah).	Markdale. Below middle.	Durham. Near top.	Flora. About middle.	Fergus. Below middle.	Rockwood. Base.	Guelph. 50 ft. up.	Hespeler. Upper beds.	Galt. Upper beds.
Anthozoa.																	
Pycnostylus guelphensis W...		× ×		×	×				×					×			× ×
elegans W...		×															×
Diplophyllum ... (Hall)...									×	×				×			
Favosites ... (Goldfuss)...	•	× ×	•		•		×	•	×	•				• ×			×
... niagarensis (Hall)...																	×
Favosites hisingeri Edwards and Haime...				×	×		×	×	×				×				
Halysites catenularia (Linnaeus)...	×																
Halysites catenularia microporus (Whitfield).																	
Annelida.																	
Cornulites arcuatus Conrad...																	
Brachiopoda.																	
Rhinobolus galtensis (Billings)...							×										×
Trimerella grandis Billings...	×	×	×	×	×	×	×	×	× ×	×	×	×	×		×		×
Stropheodonta profunda (Hall)...		•			•												
Leptaena rhomboidalis (Wilckens)...		• ×	×	×	×			×	× ×	×			×	• ×	×		× ×
... elegantula (Dalman)...																	
Anastrophia internascens Hall...			×		×		×		×								
Conchidium ... le Hall...																	
Pentamerus oblongus Sowerby...			×	×	×	×	×	×	×	×	×	×	×		×		×
Clorinda ventricosa (Hall)...																	
Rhynchotreta cuneata americana (Hall)...																	
Camarotoechia (Stegerhynchus) neglecta (Hall)...																	
Camarotoechia ...																	
Wilsonia saffordi (Hall)...																	
Atrypa reticularis (Linnaeus)...		×		×	× ×				× × •								
Spirifer radiatus (Sowerby)...																	
Spirifer crispus (Hisinger)...																	
Spirifer eudora (Hall)...																	×

Distribution of the Guelph Fauna—Continued.

Genera and species.	West end Fitzwilliam Island. Lower.	Middle island. Middle.	Cove and Fitzio Islands. Upper.	Cape Hurd. High up.	Tobermory. Upper.	Halfway rock. 100 ft. up.	Stokes bay. Basal beds.	Fishing islands. Lower.	Zinc prospect. Near Wiarton (Basal).	Markdale. Below middle.	Durham. Near top.	Elora. About middle.	Fergus. Below middle.	Rockwood. Base.	Guelph. 50 ft. up.	Hespeler. Upper beds.	Galt. Upper beds.
Brachiopoda—Con.																	
Whitfieldella nitida (Hall).	x				x x x												x
Whitfieldella nitida oblata (Hall).																	x
Whitfieldella hyale (Billings).					x												x
Pteropoda.																	
Amphicelia leidyi Hall.				x			x x		x x x x		x						
Megalomis canadensis Hall.				x			x		x x x								
Matheria recta (Hall).					x				x		x						
Mytilarca acutirostra (Hall).																	
Pterinea brisa Hall.												x					
Pterinea occidentalis Whiteaves.							x										
Anodontopsis concinna Whiteaves.	x			x x x x x x x						x				x		
Gastropoda.																	
Tremanotus alpheus (Hall).																	x
Eccyliomphalus circinatus (Whiteaves).											x						x
Euomphalopterus valeria (Billings).										x							x
Liospira perlata (Hall).																	x
Ecotomaria galtensis (Billings).				x							x		x				x
Coelocaulus bivittatus (Hall).				x								x	x	x			x
Coelocaulus longispira (Hall).				x			x							x			x
Coelocaulus macrospira (Hall).				x										x			x
Coe... ac turritiformis (Hall).				x							x						x
" constricta (Whiteaves).				x													x
...ina daphne (Billings).	x		x	x							x				x	x	x
Lox... pica hespelerensis (Whiteaves).				x										x			x
Lophospira bispiralis (Hall).				x							x	x				x	x
Lophospira mylitta (Billings).	x			x													x
Euomphalus galtensis Whiteaves.	x			x							x	x				x	x

Species		
Pelecunita crenulata (Whiteaves)	w	
Pelecunita? sulcata (Hall)	w	a
Pycnamphalus solarioides (Hall)	w	
Holopea guelphensis Billings		
Conularia.		
Conularia laqueata Conrad		w
Cephalopoda.		
Orthoceras brucensis williams sp. nov		
(Cycloceras? sp.		
Orthoceras unionense (Worthen)		
Dawsonoceras americanum (Foord)		w
Kionoceras darwini (Billings)	w	
Protokionoceras crebesceus (Hall)	w	
?Sphyridoceras toxistum (Hall)		
?Sphyridoceras desplainense McChesney	w	
Cyrtoctonaceras orodes (Billings)	w	
Melonoceras arricanneratum (Hall)		
Poterioceras n. sp.		
Poterioceras saurideus Clarke and Ruedemann	w	
Trilobita.		
Arctinurus boltoni (Bigsby)		
Calymene niagarensis Hall		
Cheirurus niagarensis (Hall)		

w-reported by Whiteaves but not found by the author, a - near locality named.

[1]This is probably Hall's Orthoceras inhabratum? Pal. N.Y., vol. II, pl. 62. Some specimens measure 6 inches in dia

Revision by Chadwick.

REVISED LIST OF SPECIES.

The following is Whiteaves' list of the fossils of the Guelph ... 1 ati of Ontario, with the doubtful species o 1 itted and the generic nam corrected according to Bassler's index of Ordovician and Silurian fossi Localities are fro 1 Whiteaves' earlier lists.

ANTHOZOA.
Pycnostylus guelphensis, Hespeler, Elora, Durham.
Pycnostylus elegans Whiteaves, Hespeler, Durham.
Favosites hisingeri, Edwards and Haime, Galt, Hespeler, Elora, Durham.
Halysites catenularia, Linnaeus, Guelph, Elora, Hespeler, Durham.
Halysites compactus, Rominger, Galt, Elora.

HYDROMEDUSAE.
Note. Parks list is given below instead of that given by Whiteaves.

BRACHIOPODA.
Trimerella grandis Billings, Galt, Elora, Hespeler.
Trimerella acuminata Billings, Galt, Elora, Hespeler.
Trimerella ohioensis Meek, Durham
Trimerella billingsi Dall, Hespeler.
Trimerella dalli Davidson and King, Hespeler.
Rhinobolus galtensis (Billings), Galt, Elora, Durham.
Monomorella prisca Billings, Hespeler, Elora.
Monomorella orbicularis Billings, Hespeler.
Monomorella ovata Whiteaves, Durham.
Monomorella ovata lata, Whiteaves, Durham.
Monomorella durhamensis, Whiteaves, Durham.
Pentamerus oblongus Sowerby, Durham.
Conchidium occidentale Hall, Galt, Guelph, Hespeler, Elora, Durham.
Clorinda ventricosa (Hall), Hespeler, Elora.
Camarotœchia pisa (Hall and Whitfield), Hespeler and Elora.
Atrypa reticularis (Linnaeus), Hespeler.
Spirifer radiatus (Sowerby), Durham, Elora.
Spirifer crispus (Hisinger), Durham.
Whitfieldella hyale (Billings), Galt, Hespeler, Elora.

PELECYPODA.
Megalomus canadensis Hall, Galt, Guelph, Hespeler, Elora, Durham, Bellwood.
Megalomus compressus, Nicholson and Hinde, Hespeler.
Amphicœlia neglecta (McChesney), Elora.
Goniophora crassa Whiteaves, Durham.
Illionia canadensis Billings, Elora and Hespeler.
Illionia ? costulata Whiteaves, Elora and Durham.
Anodontopsis concinna Whiteaves, Galt and Durham.
Prolucina galtensis Whiteaves, Galt, Durham.

GASTROPODA.
Scenella conica Whiteaves, Durham.
Archinacella canadensis (Whiteaves), Hespeler.
Tremanotus angustatus (Hall) = *T. alpheus?* Galt, Guelph, Hespeler.
Euomphalopterus valeria (Billings), Durham, Bellwood.
Euomphalopterus halei (Hall), Durham.
Euomphalopterus elora (Billings), Galt.
Lophospira contradi (Hall), Elora.
Lophospira zanthippe (Billings), Galt.
Lophospira mylitta (Billings), Elora.
Lophospira hespelerensis (Whiteaves), Hespeler.
Lophospira guelphica Whiteaves, Durham.
Loxoplocus solidus (Whiteaves), Hespeler, Elora, Durham.

[1]Palæozoic fossils, pt. IV, pp. 327-345.

GASTROPODA. Con.
   Phanerotrema occidens (Hall), Elora.
   Liospira perlata (Hall), Galt.
   Eotomaria galtensis (Billings), Galt.
   Eotomaria durhamensis (Whiteaves), Durham.
   Clathrospira deiopeia (Billings), Elora.
   Hormotoma whiteavesi Clarke and Ruedemann, Galt, Hespeler, Elora.
   Cœlocaulus? vitella (Billings), Galt.
   Cœlocaulus? macrospira (Hall), Galt.
   Cœlocaulus? bivittatus (Hall), Galt, Hespeler, Elora.
   Cœlocaulus? longispira (Hall), Galt, Elora, Hespeler, Guelph, Durham.
   Cœlocaulus? turritiformis (Hall), Galt, Elora, Hespeler, Durham.
   Cœlocaulus? estella (Billings), Galt.
   Turritoma boylei (Nicholson), Elora.
   Turritoma constricta (Whiteaves), Durham.
   Murchisonia? billingsana Miller, Galt.
   Lophospira bispiralis (Hall), Galt.
   Pleurotomaria cyclostoma Whiteaves, Durham.
   Pleurotomaria townsendii, Durham.
   Euomphalus galtensis Whiteaves, Galt, Hespeler, Durham.
   Euomphalus inornatus (Whiteaves), Elora, Durham.
   Eccyliomphalus circinatus (Whiteaves), Galt, Hespeler, Elora, Durham.
   Straparollus hippolyta Billings, Galt.
   Straparollina daphne (Billings), Galt.
   Poleumita sulcata (Hall), Galt, Hespeler.
   Poleumita macrolineata (Whitfield), Elora, Durham.
   Poleumita crenulata (Whiteaves), Durham.
   Poleumita durhamensis (Whiteaves), Durham.
   Poleumita parvula (Whiteaves), Durham.
   Codonochilus striatus Whiteaves, Hespeler, Durham.
   Pycnomphalus solarioides (Hall), Galt.
   Loxonema boydi Hall, Galt.
   Holopea harmonia Billings, Galt.
   Holopea ? occidentalis Nicholson, Elora.
   Subulites compactus Whiteaves, Durham.
   Cyrtospira ventricosa (Hall), Galt.

CEPHALOPODA.
   Protokionoceras crebescens (Hall), Hespeler, Elora.
   Protokionoceras medullare (Hall), Elora.
   Orthoceras selwyni Billings, Galt.
   Kionoceras cancellatum (Hall), Elora.
   Kionoceras darwini (Billings), Hespeler, Durham.
   Dawsonoceras americanum (Foord), Hespeler, Elora.
   Septameroceras septore (Hall), Elora.
   Ascoceras townsendi Whiteaves, Durham.
   Maelonoceras arcticameratum (Hall), Galt.
   Cyclostomiceras orodes (Billings), Hespeler.
   Phragmoceras hector Billings, Hespeler.
   Phragmoceras nestor canadense Whiteaves, Hespeler, Elora, Durham.
   Phragmocerus parvum Hall and Whitfield, Durham.
   Sphyradoceras desplainense (McChesney) "Ontario".
   Discoceras graftonense Meek and Worthen, Elora, Hespeler.

CRUSTACEA.
   Leperditia balthica guelphica (Jones), Durham, Aboyne.
   Leperditia praseolus guelphica Jones, Durham.

TRILOBITA.
   Calymene niagarensis Hall, Galt.
   Cheirurus niagarensis (Hall), "Ontario".
   Illaenus aboynensis Whiteaves, Aboyne.

EURYPTERIDA.

Eurypterus (Thylopterus) boylei, Whiteaves, Elora.
Stromatoporoids of the Guelph formation in Ontario as given by Parks.[1]
Actinostroma culcana Parks, Durham.
Clathrodictyon ostiolatum Nicholson. No locality given for Guelph occurrence.
Clathrodictyon striatellum D'Orbigny, " at all the typical localities ".
Clathrodictyon fastigiatum Nicholson, Elora, Aboyne, and township of Glenelg.
Labechia durhamensis Parks, " Durham, Gait, Guelph, Elora, and throughout the
    Guelph of Ontario ".
Labechia mik.ora Parks, Elora, Gait, Durham, Glenroading.
Rosenella glenelgensis Parks, Durham, Eiora.
Stromatopora galtensis Dawson, Gait, Hespeler, Elora, Durham.
Stromatopora antiqua Nicholson and Murie, Durham?
Stromatoporella elora Parks, " at all of the well-known localities ".
Stromatoporella elora minuta Parks, Durham.
Hermatostroma guelphica Parks, Elora.

CAYUGAN GROUP.

GENERAL DESCRIPTION.

"Cayugan group" is here used to include all the Silurian sedi: ents in Ontario above the top of the Guelph dolomite. Thus in the Buffalo region, the for: ations included are, as recognized by Bassler[2], the Salina (represented by the Camillus shale), the Bertie waterlime, and the Akron dolo: ite, the latter correlated with the Cobleskill of eastern New York. In the Detroit River region the Salina and the Bass Island series are included in the Cayugan group. The Camillus shale division of the Salina is well: arked in this western region and it is probable, as will be seen below. that the Bertie and Akron dolo: ites of the east are to be correlated with the Raisin River and the Put-in-Bay dolo: ites of the Bass Island series, and conse: uently, contrary to for: er usage, this series belongs in the Cayugan group.

The lack of outcrops and good sections is the principal difficulty met with in working out the stratigraphy and correlation of the Cayugan for: a- tions. This difficulty is illustrated by the scattered occurrence of outcrops as shown on Map 1715.

SALINA FORMATION.

Description.

The Salina, as its na: e i: plies, is the salt and gypsu: -hearing for: ation of New York state and Ontario. The deposits of Salina age represent special conditions of sedi: entation, which according to Grabau and O'Connell, are probably those of an interior basin, to which the Atlantic ocean gradually gained freer access until the co: : ence: ent of the e: er- gence of the continent as a whole, which closed the Silurian period.

[1] University of Toronto studies, Geol. ser No. 4, 1917.
[2] U.S. Nat. Mus., Bull. 92, vol II. 1915, p. 3.

Camillus Shale Member.

Description. The Camillus shale is the lowest member of the Salina formation of Ontario; it rests unconformably on the Guelph dolomite and is overlain conformably by the Bertie waterlime in the east and the Bass Island series in the west.

Essentially a grey or green shale, this member contains also beds of argillaceous dolomite, large deposits of gypsum in the valley of Grand river between Paris and the vicinity of Cayuga, and extensive salt deposits in parts of New York state and in the area adjacent to lake Huron from Kincardine south to lake St. Clair and near Windsor.

The characters of this formation are illustrated in the following descriptions of sections, and Figure 4.

A drill core from north Buffalo as interpreted by Grabau gives a section of the Camillus, 387 feet thick, which includes, in ascending order: gypsum and shale, mottled and in streaks, 290 feet; compact shale, 3 feet; shale and limestone, 4 feet; dark limestone, 2 feet; drab shale and thin gypsum, 58 feet; shale and gypsum, 7 feet; white gypsum, 4 feet; shale, 1 foot; white gypsum, 12 feet; shale, 2 feet; pure white gypsum, 4 feet; overlain by the Bertie waterlime. The beds penetrated in the gypsum mine of the Alabastine Company at Caledonia are, in ascending order, as follows: 3 inches of gypsum; 4 feet of dark grey, impure dolomite; 3 feet of fine white gypsum; 10 feet of brown dolomite containing some gypsum; 7 feet of gypsum with thin, limy partings and limy accumulations; 4 feet of mixed dolomite and gypsum; 3 feet of impure dolomite; 6 feet of gypsum mixed with dolomite. The weathered exposures of shale along Grand river are grey-green in colour and generally soft and fissile.

The upper Camillus shales are represented by pea-green to grey, hackly shale in outcrops along Niagara river 100 yards north of the international bridge at Bridgeburg, Ont. Indications of gypsum are seen a little farther down stream. The best of the northern sections is that below the bridge over the branch of Saugeen river at Ayton. In ascending order the section is: 4 feet of hard, firm shale; 4 feet of hard, uneven dolomite, breaking with conchoidal fracture; 4 feet of hard, thin-bedded, buff dolomite with bituminous partings; 2 feet of hard, brown dolomite; 7 feet of firm, buff shales and thin dolomites; 11 feet of thin, fissile, green-weathering shale; 4 feet of light buff dolomite full of minute cavities.

Another section is the bank of a branch of the Saugeen river, on lot 19, concession VI, Normanby township, Grey county, is as follows: 3 feet shale; 10 feet hard, grey dolomite; 12 feet soft shale; 15 feet of nodular, green shale; 2 feet green shale containing red patches, probably originally all red; 2 feet of buff dolomite (calcilutite); 2 feet of green-grey, impure dolomite. This is the only locality known to the author where red shale occurs in a Camillus shale outcrop. The red shale was probably more extensive than is shown, the red beds being largely deoxidized by surface water containing humic acid derived from plants.

In the west bank of Saugeen river below Walkerton, at the crossing of the line between concessions III and IV, about 16 feet of nodular green shale is overlain by 20 feet of buff dolomite, belonging to the Bertie-Akron division.

The total thickness of the Camillus shale at Delhi, Norfolk county, is 310 feet; near London the thickness, including 170 feet of salt, is 750 feet; in Dover township, near lake St. Clair, the thickness varies between 550 and 745 feet, the least thickness being where the underlying Guelph dolo ite is thickest and *vice versa*.

The varying thickness of the shale an l the irregularities in the underlying for ation are suggestive of warping, uplift, and erosion at the end of Guelph ti e, as described by Grabau.

Origin of Camillus Shale.

Grabau has ade a careful study of the origin of the Camillus shale, with the contained salt and gypsu , and concludes[1]:

"At the close of the Niagara period, there appears to have been an elevation of the c ent which converted . . . . . the greater part of the interior Siluric sea into a vast partially or entirely enclosed basin. This elevation appears to have been acco panied by climatic desiccation which brought about a rapid evaporation of the waters and a consequent increase in salinity. Thus this great interior water body was changed fro a richly peopled mediterranean, to a lifeless body of intensely saline water, a veritable Dead sea. . . . . The elastic strata of the Salina series were probably derived fro the destruction of the sediment which were formed during the early periods of the Siluric and during preceding periods . . . . . they may in fact be regarded as c solidated argillo-calcareous uds derived fro older li estones and shales."

The salt deposits of New York state and the uch greater deposit of the Michigan basin, whose eastern ri includes the Lake Huron-Lake St. Clair region of Ontario, were evidently the result of extre e desiccation in isolated basins. Deposition of gypsu took place as its saturation point was reached, and later the salt was deposited. These depositions alternated, responding to ore or less periodic inflows of water and excess of evaporation. It has been deter ined that salt co prises about 77 8 per cent of the solids contained in sea water and calciu sulphate co prises 3 6 per cent. The latter, however, begins to precipitate when 80 per cent of the water is evaporated, salt not being deposited until 93 per cent of the water is evaporated. It is clear that in cases of extre e evaporation the deposits of calciu sulphate (gypsum and anhydrite), because of its s all content in sea water, will for very insignificant deposits co pared with that of the salt. If total desiccation took place the deposits would be in the ratio of 3 6 parts by weight of anhydrite to 77 8 parts by weight of salt. If just sufficient water were added fro ti e to ti e to the evaporating basin to prevent desiccation reaching the stage necessary to deposit salt, calciu sulphate would continue to be deposited and i portant beds of anhydrite and gypsu would be for ed, as was the case in parts of New York state and the Paris-Cayuga section of the Grand valley in Ontario.

BERTIE WATERLIME FORMATION.

Description.

The Bertie waterlime for ation is co posed of thin-bedded dolo ites, jointed waterlimes, and interbedded, dark grey shale. The lower contact

[1] N.Y. State Mus. Bull. 45. pp. 127-8.

is transitional from the Ca illus shale, and the upper contact with the Akron dolomite is marked by a lithologic change, which is not everywhere easily detected. As described by Grabau, the Bertie is represented in a drill core from north Buffalo, in ascending order as follows: 13 feet of shale and ce ent rock at the base, 5 feet of fairly pure cement rock, and 25 feet of shale and ce ent rock at the top.

Chadwick[1] has divided the upper part of the Ca agan group in western New York as follows (all but the Akron are divisio s of the Bertie for a tion):

	Feet.
Akron dolomite (Grabau), subcrystalline, grey to brownish, with *Cyathophyllum hydraulicum*	12 or less
Buffalo cement bed carrying eurypterids	6
Scajaquada dar shales and blocky waterlimes, at base the Bridgeburg horizon with eurypterids	8
Falkirk dolomite, brownish and bituminous, below massive and often producing waterfalls; a marine fauna, but eurypterids at base	30
Oatka beds, dark grey and shaly, with a blocky waterlime at base carrying eurypterids	20 or less.

In the criticis of this memoi er, Chad k has omitted the ter Buffalo and used in its place amsville.

The shales and waterlime along t ul Trunk tra s at Bridge burg, probably belong to the upper the Bertie ajaquada in cluding Bridgeburg and top of Falkir te a d the sr le just south of the international bridge is probably base the Bertie division.

At a quarry north of Ridgeway, o ence nea X. ertie town ship, Welland county, the following pa ction of Bertie beds occurs: 7 feet of thinly la inated, brown dol with bit minous partings at the base of the section (Falkirk); 4 f of light grey, jointed olomites (Bridgeburg); 6 feet of hard, slate-grey shale, w h B ostend fragments, at the top (Scajaquada); 5 feet of wa erlime ville over an by 6 feet of Akron dolo ite.

About one-third of a ile due west of Rid point at the road forks, heavy beds of dolo ite (Falkirk) c tain num s specimens of *Whit fieldella sulcata* and ostracoda. These ds c erl a out 16 feet of thinner beds which are probably lower Bert Scajaq eda along the Ridge road" south of the above-mentioned exposur h base of the Akron is about 13 feet below the Oriska sandston being the thickest Akron section known in the vicin

At the Municipal quarry, sou east of B a d on he west side of Grand river, not far from Dunnville, the exp d ertie section consists in ascending order, of: 7½ feet of hard, silice n ous shale with dark partings; one-half foot of hard, nodular s t of assive, hard, light grey, fine-grained dolo ite (calcilutite con *Schuchertella hydra ulica*? near the top; 2 feet of hard, grey shale 7 feet of la inated and mottled, blocky dolo ite (agnesian dolomite, probably Akron) overlain above a decided erosion unconform , by lense of Oriskany sandstone.

From Dunnville west and north few exposures of Cayugan rock occur and the Bertie and Akron beds have not been differentiated. The exposures

---

[1] Bull. Geol. Soc. Am., No. 1, vol. 28, p. 173.

will be described under Akron dolomite. It seems probable that in Kent and Essex counties the extension of the Bertie waterlime is to be correlated in age with the Put-in-Bay dolomite and the underlying Tymochtee shale and Greenfield dolomite (see below under Bass Island series).

Fauna.

The muddy, calcareous Bertie sediments contain a rich *Eurypterid* fauna, but very little else. As given by Ruedemann the fauna of the Bertie (Williamsville) is as follows:

Inocaulis vaspureigri (Grote and Pitt).
Lingula testatrix Ruedemann.
Diaphorostoma sp.
Hercynella pat ?formis O'Connell
H.   bu ba nsis O'Connell.
Hormotoma gregaria Ruedemann.
Orthoceras vicinus Ruedemann.
Phragmoceras aceula Ruedemann.

Clarke and Ruedemann list the following eurypterids[1] fro 1 the Bertie waterlime.

Dolichopterus macrocheirus Hall.
D.   siluriceps Clarke and Ruedemann.
D.   ? testudineus "   "
Eurypterus dekayi Hall.
E.   lacustris Hall.
E.   lacustris var. pachycheilus Hall.
E.   pustulosus Hall.
E.[2]   remipes Dekay.
E.   scorpionis Grote and Pitt.
Pterygotus buffaloensis Pohlman.
P.   cobbi Hall var. *juvenis* Clarke and Ruedemann.
P.   grandis (Pohlman).
P.   macrophthalmus Hall.

Chadwick by correspondence gives the following as occurring in th Falkirk division of the Bertie:

Whitfieldella sulcata (Vanuxem).
Gomphoceras osculum Ruedemann.
Leperditia cf scalaris (Jones).

Origin of the Bertie Waterlime.

According to Miss O'Connell, who has made a special study of "the Bertie waterlime fro 1 the standpoint of "the habitat of the Eurypterida" the deposits at Buffalo and at Herkimer, New York, both of which carr eurypterid faunas, are best explained as deltas of low-lying, meanderin rivers flowing fro 1 the north. The marine organis 1 s are supposed ' have been transported upstrea 1 by tidal action. On page 117, we re[3] "The only available source of the li 1 e in the Bertie is fro 1 the mur derived by the erosion of an older 1 agnesian limestone, the Niagara

[1] N.Y. State, Mem. 14, vol. I, 1912, p. 431.
[2] See Plate XXVII.
[3] See Plate XXVII.
[4] Bull. Buffalo Soc. Nat. Sc , 1916.

or in so i e cases, perhaps, the Trenton. Where the Bertie is eurypterid-
bearing, the ro k was evidently deposited above sea-level, as a river flood-
plain and subaerial delta deposit. Southward and laterally the subaqueous
part of the delta carries few or no eurypterid remains, but more marine
organis i s. That the Bertie eurypterids lived in the river is thus indicated . . . . . "

An entirely different view of the case is taken by Ruedemann[1] who,
after discussing O'Connell's arguments and refuting the suggestion that
the eurypterid fragments had drifted into their present position, su i s
up as follows:

" The *Buthotrephis* has been proved in this paper to be a good sessile
graptolite of the genus *Inocaulis*. It occurs in the waterlime in splendidly
preserved colonial stocks, which cannot have drifted any distance. The
Lingulas and *Hormotoma*, occur in certain layers in such immense quantities that they are certainly in their proper surroundings where found
. . . . . At any rate there is not in the whole assemblage of the Bertie
waterlime a single genus that could be considered as indicative of a
freshwater fauna, and all these marine for i s are found on the same slabs
with the eurypterids, so that an alternation of marine and freshwater
conditions cannot be assumed either."

The above summarizes the divergent opinions as to the probable
origin of the typical Bertie beds where recognized by faunal evidence.
Elsewhere, they have been determined by stratigraphic and lithologic
evidence and appear to be sediments derived from older rocks laid down
in a shallow sea. The water was probably too salty and too muddy
to favour the development of diversified life.

AKRON (DOLOMITE) FORMATION.

Description.

In the cement quarry at Buffalo about 7 feet of light grey, mottled,
somewhat nodular dolomite rests on the Bertie waterlime, and is terminated above by a decided erosion surface upon which lie lenses of Oriskany
sandstone and beds of Onondaga limestone. This is the Akron dolomite.
Small cavities are common in the Akron, which has in part wavy bedding,
and contains *Cyathophyllum hydraulicum* and *Schuchertella hydraulica*
(see below).

The Akron dolomite was named by Grabau[2] fro i its occurrence at
Akron Falls, east of Falkirk, New York. Here, as seen by the author,
the Akron, which has wavy bedding, is about 5 feet thick, light grey in
colour, somewhat nodular, fine-grained, and weathers into a pitted surface.
The underlying bed of Bertie waterlime was formerly burnt to natural
rock cement. The overlying for i ation is the Onondaga, but the contact
is even and the disconformity is not evident.

A good section of the Akron dolomite occurs at the quarry near
the schoolhouse north of Ridgeway, lot 1, concession X, Bertie township,
Welland county. Here the total thickness is about 7 feet. The characters
are about as at Buffalo and *Schuchertella hydraulica* and *Cyathophyllum*

[1] Ibid. p. 115.
[2] Bull. Geol. Soc. Am., vol. 19, p. 544.

hydraulicum are found. A similar section occurs at a quarry on lot 8, concession XIII, Bertie, where Onondaga limestone overlies the Akron.

At Byng near Dunnville, in the Municipal road-metal quarry, 5 to 7 feet of Akron dolomite is terminated above by an irregular erosion surface upon which lie lenses of Oriskany sandstone. The Akron consists of thin-bedded, blocky calcilutite of mottled grey colour, and rests on hard, grey shale. At the John Weber quarry near Byng, on lot 16, concession I south, Dunn township, the Akron appears to be 15 feet or more thick, but the boundary between the Akron and the Bertie is not easily drawn.

The outcrops west of Dunnville, not easily divided into Akron and Bertie, are described below. As will be seen on Maps 1714 and 1715, a number of outcrops of the upper Cayugan dolomites occur in the vicinity of Cayuga west of Grand river. The best and most representative section is that north of the road in the river bank, on the fronts of lots 3, 4, and 5, River road, Cayuga, South, township, Haldimand county. In ascending order the section is: Camillus shale consisting of 12 feet of light grey, thin-bedded dolomite, and 8 feet of grey shale including a 1-foot bed of waterlime near the top; Bertie-Akron consisting of a 3-foot massive bed of light grey dolomite, and 20 feet of light grey, blocky dolomite, overlain by cherty Onondaga dolomite.

At Springvale, an 8-foot section is exposed in a quarry, the lower 5 feet of the beds being thin and blocky and the upper 3 feet being one massive bed. The colour is light buff. This stone was burnt for lime by Mr. Winger, who says that it burnt to a strong, white lime and that it was used in the building of the convent at Niagara Falls, New York. The waterlime bed is said to lie just below the exposed rock of the quarry. A short distance southwest of Springvale, the Springvale (Devonian) sandstone occurs, limiting the Cayugan above. Other small exposures similar to the above occur south of Springvale.

About 2 miles east of Hagersville, at a cut on the Michigan Central railway near where two roads cross, 8 feet of thin-bedded, jointed, cream-coloured, fine-grained dolomite is terminated upward by an erosion surface on which rests cherty Onondaga limestone.

At Innerkip, just south of the station, stone was formerly taken from a quarry and burnt to quicklime by the aid of natural gas which was struck on the quarry site. The face of rock now exposed above water is 8 feet high, consisting of light buff, fine-grained dolomite. The lowest bed, which is very fine-grained and breaks with conchoidal fracture, is 3 feet thick and is overlain by a thin bed of green, glauconitic, dolomitic, very fine-grained sandstone, 1½ inches or less in thickness. The beds above are thinner and weather into nodules. The main sandstone bed lies on an irregular erosion surface, and streaks of sand occur 6 inches lower down and also in cracks at a slightly lower horizon. The lime burnt from the stone in this quarry was grey.

Two specimens of *Whifieldella prosseri* Grabau were found just above the sandstone. These consist of moulds of the interior of the shell and the sinus in the pedicle valve is not well marked. The specimens compare closely in size and other characters with Grabau's illustrations and descriptions. Had the fossils occurred below the sandstone it might be assumed that these beds were the top of the Silurian. The sand is probably, how-

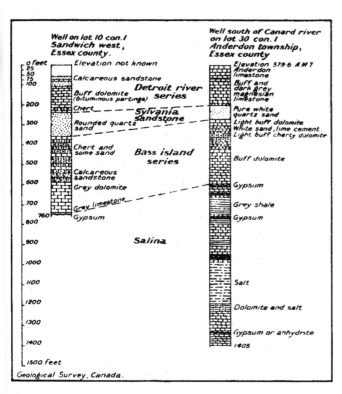

Well on lot 10 con.1
Sandwich west,
Essex county.

Well south of Canard river
on lot 30 con.1
Anderdon township,
Essex county

Elevation not known

Calcareous sandstone

Buff dolomite
(bituminous partings)

Chert

Rounded quartz
sand

Chert and
some sand

Calcareous
sandstone

Grey dolomite

Grey limestone

Gypsum

Salina

Detroit river
series

Sylvania
sandstone

Bass island
series

Elevation 579·6 A M T
Anderdon
limestone

Buff and
dark grey
magnesian
limestone

Pure white
quartz sand

Light buff dolomite

White sand, lime cement.

Light buff cherty dolomite

Buff dolomite

Gypsum

Grey shale

Gypsum

Salt

Dolomite and salt

Gypsum or anhydrite

1405

Geological Survey, Canada.

Figure 5. Diagram showing formations near Detroit river as indicated
by two wells drilled by the Solvay Process Company.

To accompany Memoir by M.Y.Williams

ever, a forerunner of the Sylvania which has been deposited on a li estone surface scoured by shore currents during a partial retreat of the sea. The presence of *Whitfieldella prosseri* in these beds correlates the with the Raisin River dolo ite of Ohio and Michigan. This outcrop forms a link between the Akron dolo ite and the Raisin River dolo ite, which as explained below are thought to be of the sa e age.

As mentioned above under "Camilius shale", about 6 feet of fine-grained, crea -coloured dolo ite is exposed in the bed of a s all tributary entering Saugeen river fro the west just north of Walkerton. Along the Canadian Pacific railway, just south of Walkerton, buff dolo ite with bitu inous partings contains *Leperditias* rese bling *altoides* Grabau and *Whitfieldella sulcata* (Vanuxem)? In the east bank of the Saugeen, at the crossing of the boundary between concessions III and IV, Brant county, about 20 feet of buff dolo ite, fairly thick-bedded in the lower third and thinner-bedded above, overlies nodular, green shale. The dolo ite is dense and fine-grained, suggesting the characters of lithographic stone.

The best northern section of the Bertie-Akron is that exposed along a s all un apped tributary entering Saugeen river fro the west just south of the line between concessions VII and VIII, Brant township, and about 2½ miles north of Walkerton. The section is as follows, in ascending order fro the bed of the Saugeen up: 4 feet of buff dolo ite in 1 to 2-inch beds; a covered interval of about 3 feet; 5 feet of thin-bedded, buff dolo ite; 10 feet of buff dolo ite in 1 to 2-foot beds, 4-foot assive bed, fine-grained and hard, for erly uarried and burnt to grey li e at nearby kilns; 1 foot of soft, yellow-weathering shale; 12 feet of dolo ite in beds 1 to 2 feet thick. S all exposures of rock referred to the Bertie-Akron division occur on the Teeswater river about 1 ile below Pinkerton, and on a tributary of the Teeswater one-half ile east of Bradley.

Fauna.

The fossils contained in the Akron dolo ite of New York as given by Grabau[1] and corrected by Chadwick, are as follows:

PLANTS. *Nematophycus crassus* (Penhallow) horizon doubtful.
ANTHOZOA. *Cyathophyllum hydraulicum* Simpson.
BRACHIOPODA. *Schuchertella interstriata* (Hall).
Spirifer eriensis (Grabau)?

BASS ISLAND SERIES.

Definition and General Description.

Grabau has applied the na e Bass Island series to the dolo ites and shales of Ohio and Michigan, lying between the top of the illus shale as recognized in that region and the Sylvania sandstone. His subdivisions are, in ascending order: the Greenfield dolo ite; Tymochtee shales; Put-in-Bay dolo ite; and Raisin River dolo ite. These divisions were studied at widely separated localities and the boundaries and relations are hence not very well known. As seen in Figure 5, the thickness of the

[1] N.Y. Mus., Mem. 45, pp. 233-237.

Bass Island series along Detroit river is fro 330 to 375 feet. The follow ing descriptions are fro Grabau and Sherzer[1].

The Greenfield dolo ite is ostly thin-bedded, drab-coloured o fresh fractures, and oxidizes to a yellowish shade. Its greatest thicknes at Greenfield, Ohio, is 100 feet. The principal fossils of the Greenfiel dolo ite are:

Schuchertella hydraulica (Whitfield).
Hindella (?) *(Greenfieldia) whitfieldi* Grabau.
Hindella (?) *(Greenfieldia) rostralis* Grabau.
Pentamerus pæsoris Whitfield ?
    (horizon doubtful).
Whitfieldella subsulcata Grabau,
Whitfieldella rotundata (Whitfield).
Rhynchospira praeformosa Grabau.
Camarotœchia hydraulica (Whitfield).
Leperditia ohioensis Bassler.
Leperditia angulifera Whitfield.
Sphaerococcites? glomeratus Grabau.

At their typical exposure, the Ty ochtee shales are 24 feet thick an consist of tough, thin-bedded, bitu inous shale. The contained fossil are *Leperditia alta* and a species of ?*Modiolopsis*.

The Put-in-Bay dolo ite consists of ore than 60 feet of assive brecciated, dolo ites interbedded with thin-bedded dolo ites. Th colour ranges fro blue to grey and crea . The co onest fossils foun in the upper beds are *Goniophora dubia* (Hall) and *Spirifer ohioensi* (Grabau). The less co on species are *Leperditia alta* (Conrad) an *Eurypterus micropthalmus*. The beds of this division are not known i contact with the Ty ochtee shales or the Raisin River dolo ite and thei position and extent are hence somewhat proble atical.

The Raisin River dolo ite consists of grey-drab, agnesian . lutites and oolites and is about 200 feet thick. The following fossils w obtained between 87 and 138 feet below the Sylvania sandstone in tl Detroit salt shaft.

PLANTÆ.   *Sphaerococcites? glomeratus* Grabau.
BRACHIOPODA.   *Pholidops* cf *orata* Hall.
               Schuchertella hydraulica (Whitfield).
               Schuchertella interstriata (Hall)
               Whitfieldella piesseri Grabau.
               Camarotœchia sp.
PELECYPODA.   *Pterinea lanii* Grabau,
               Goniophora dubia (Hall).
               Tellinomya sp.
               Modiomorpha sp.
GASTROPODA.   *Ectomaria minuta* (Hall).
               Holopea three sp.
               Loxonema sp.
CEPHALOPODA. *Cyclostomiceras orodes* (Billings).
OSTRACODA.   *Kloedenia monroensis* Grabau.

Occurrences.

No outcrops of upper Cayugan rocks that can be directly correlat with the Bass Island series occur in Ontario, excepting the outcrop

Innerkip described above under Akron dolo 1 it , and rocks for 1 ing the t'anadian islands and reefs west of Peice island. These islands are only a short distance fro 1 Bass islands, fro 1 whence the series 1a 1 e was obtained, and are described below fro 1 a brief study 1 ade of the 1 by the author.

Hen island, which is nearest Pelee island, is a flat, pa1k-like island of 4 or 5 acres ex:ent, fitted out with club houses, a dock, etc. Its highest elevation is between 10 and 15 feet above the lake. Fine trees consisting of elm, maple, and hackberry furnish abundant shade. As 1 ight be expected fro 1 its exposed position, Hen island is underlain by resistant rock—a dense, tine-grained dolo 1 ite The best exposure is on the northwest side where the following section is exposed in ascending order: 3 feet of 1assive, uneven weathering, tine-grained, buff dolo1ite containing nu 1 erous speci 1 ens of *Whitfieldella prosseri* Grabau and a few speci 1 ens of *Rhynchospira praformosa* Grabau?; 1 foot of light grey, la 1 inated dolo 1 ite; 2 feet of 1 assive, buff, fine-grained dolo 1 ite; 2 feet of 1 assive, light brown dolo 1 ite containing 1 inch or 1 ore of sand 1 ingled with the dolo 1 ite near its base. This sand, which 1 akes up about half the 1 aterial of the thin bed, weathers into relief, and shows the characters of the Sylvania sand. At the top of the section is a 2-foot bed of dense, tine-grained dolo 1 ite containing vugs of celestite (?) and weathering into a scraggy rock with nu 1 erous irregular spaces. The strike of the bedding at this locality is about east, 20 degrees south and dip south 6 degrees. Thus the strata exposed at the extre 1 e eastern end of the island are lower in the section than those just described, about 5 feet 1 ore of the lower massive bed being exposed.

Big and Little Chicken islands are 1 ere reefs washed by the waves except in the cal 1 est weather.

East Sister, the largest of the group of islands, is low and flat and was for 1 erly cultivated. Club houses are situated near the southeast end and peach trees planted so 1 e ti 1 e before were struggling with weeds in Septe 1 - ber, 1914. A fringe of hackberry, 1 aple, and wild grape su1rounds the island. A gravel bar about 5 feet high is piled up along the south side, the gravel being in general under 2 inches in dia 1 eter and consisting of about 75 per cent dolo 1 ite and 25 per cent granite, etc.

At the western end of the island a s 1 all syncline occurs with direction of axis about north 40 degrees west. The dips toward the 1 iddle are about 2 to 3 degrees and the width of the syncline is only about 50 feet. Exposures to the east show buff dolo 1 ite overlying oölitic dolo 1 ite which occupies troughs 2 feet wide and 1 foot deep in the underlying rock, along an exposed length of 4 feet. This oölite corresponds closely with the description of the Raisin River oölites given by Grabau.[1] Farther west, thinbedded, grey dolo 1 ite underlies oölitic dolo 1 ite about 2 feet in thickness. Below, to the east, dark massive rock is exposed containing numerous specimens of *Whitfieldella prosseri* Grabau and a few specimens (one observed) of *Spirifer ohioensis* Grabau. The extre 1 e eastern end of the island consists of a do 1 e of rock, about 50 feet wide, extending nearly east and west. Glacial striae here run about 5 degrees north of west. Probably not 1 ore than 25 feet of strata are exposed on the island. the exposures at the west end representing the top and those at the 1 ast end representing the botto 1 of the section.

Grabau and Sherzer, p. 36.

57237—7½

North Harbour, the smallest of the islands, is about 300 yards long by 50 yards wide, and is covered with hackberry and maple trees. The shores are rocky except at the east end where the rock is covered with gravel. About 6 feet of beds are represented on the island, consisting of: light grey dolomite in 6-inch beds extending below the water; 2 feet of arenaceous dolomite with an irregular upper surface; and a massive 3-foot bed of dense, grey dolomite at the top. The remains of a fossil were seen in the upper bed, but no identification could be made. The beds lie nearly flat. The water on the south side is deep to the shore, that on the north side being shallower. Glacial striæ on the north side run about 8 degrees north of west.

Middle Sister island is well wooded with maple, hackberry, and elm and has long been used as a fishing station. At the north end, and on the southwest, the rock is overlain by gravel. The best rock exposures are on the northwest side where a 10-foot section occurs, consisting of 6 feet of thin-bedded, buff dolomite and 4 feet of fine-grained, dense, buff grey dolomite at the top, which is irregularly bedded and contains fragments of *Whitfieldella prosseri* Grabau. Strata at the southwest end strike east 20 degrees south and dip south 6 degrees. Glacial striæ at the southwest end run due west.

From the fossil evidence, there is no doubt of the Cayugan age of the dolomites of the islands described and from the presence of a sand streak in the dolomite of Hen island, it seems probable that the Sylvania sandstone itself was not much higher up. The close similarity in the lithology and fossils of all the islands suggests strongly that the same horizons are exposed on them. This supposition is further supported by the fact that the islands are probably on the strike of the same beds as a slight bending to the south of the western extension of the strike of the beds of the more easterly islands would include Middle Sister island. In other words, these islands are a remnant of the upturned edges of strata which alone in the whole formation were hard enough to resist glacial, wave, and all other erosion sufficiently to remain above the lake-level to the present time.

Of the fossils, *Spirifer ohioensis* is found in the upper part of the Put-in-Bay dolomite at Peach point, Put-in-Bay island; *Whitfieldella prosseri* is characteristic of the Raisin River dolomite, being found throughout the section of this formation in the Detroit salt shaft, and also at New Port, and Monroe, Michigan, and at Peach point, Put-in-Bay, where it is thought to mark the base of the Raisin River dolomite. *Rhynchospira præformosa* doubtfully determined for Hen island, is described from the Greenfield dolomite. The positive evidence thus points definitely to the Raisin River age and correlation of the strata exposed on the islands. The oölite found on East Sister island is apparently identical with the Raisin River oölites described by Grabau and is a further indication of identity with the Raisin River dolomite. The sandstone of Hen island suggests that the beds are near the base of the Sylvania sandstone and hence near the top of the Raisin River formation. *Spirifer ohioensis*, on the other hand, suggests correlation with the lower portion of the Raisin River dolomite, for this species is found in the beds at Peach point which Grabau placed at the top of the Put-in-Bay dolomite, but which may belong to the base of the Raisin River, as *Whitfieldella prosseri* is reported from the same place and the latter is exclusively a Raisin River species. If Hen island represents strata near the top of the Raisin River dolomite, there is plenty of room for the

Detroit River series, known to be thin east of lake St. Clair, to outcrop between this island and Pelee island which consists of rocks of Onondaga age. It may be noted that the distance here is less than that between Put-in-Bay island and Middle island (Onondaga), within which distance the Raisin River as well as the Detroit River series must outcrop.

Correlation of Bertie and Akron Formations with the Bass Island Series.

It has been shown above that Raisin River dolomite outcrops at Innerkip, and this locality is a connecting link between the regions occupied by well-defined Akron and Bertie dolomites and the Bass island series. Although positive proof is lacking, there is a strong indication that the Akron and the Raisin River dolomites are to be correlated at least in part. The presence of *Schuchertella interstriata* in both formations supports this correlation and the somewhat close relationship of the *Whitfieldellas* of these formations is suggestive. Again, the presence of *Leperditia alta* (Conrad) and *Eurypterus micropthalmus* Hall, in the upper beds of Put-in-Bay (which may yet have to be considered basal Raisin River), are clear evidence of Manlius affinities, since both these species occur in the Manlius of New York.

If the Akron is to be correlated with the Raisin River formation, including the upper beds in Put-in-Bay island, it would be natural to expect that the Bertie might be correlated with the Put-in-Bay dolomite if this lies just below, and the 12 foot of waterlime found in the lower third of the section at Put-in-Bay is very suggestive of the Bertie beds. There is a strong probability that the Akron dolomite of New York, as stated above, is to be correlated with the Raisin River dolomite, and that the Bertie waterlime may be correlated in part with the Put-in-Bay dolomite (if the latter has the relations assigned to it by Grabau) and in part with the Tymochtee shale and Greenfield dolomite. That the basin of sedimentation in Michigan and Ohio was deeper than that in New York, is shown by the thicker dolomite deposits.

Professor G. H. Chadwick disagrees with the above conclusions and offers the following alternative correlation.

Believing that the Put-in-Bay dolomite is above the Raisin River dolomite rather than below it as described by Grabau, and accepting the author's correlation of the Akron dolomite with the Raisin River dolomite, he correlates the Put-in-Bay dolomite with the lower Manlius (= Cobleskill); the Tymochtee shale with the upper Bertie (his Williamsville and Seajaquada divisions), and the Greenfield dolomite with the lower Bertie) (his Falkirk and perhaps Oatka divisions).

His table is as follows:

Detroit region.	Western New York.	
Put-in-Bay dolomite................	.... Lower Manlius = Cobleskill dolomite.	
....	?"Roundout" waterlime	
Raisin River dolomite.................	Akron dolomite.	
Tymochtee shale....	Williamsville waterlime and shale.	
	Seajaquada limy shale.	
Greenfield dolomite...	Falkirk dolomite	Bertie.
	?Oatka dolomite.	
	Salina formation—shale, etc.	

This contention is supported by the presence of the Maulins species, *Leperditia alta*, and *Eurypterus microthalmus* in the upper beds on Put-in-Bay island, but on the other hand *Spirifer ohioensis* and *Whitfieldella prosseri*, as shown above, suggest the correlation of these beds with the lower part of the Raisin River dolomite.

It appears evident that no final correlation can be made until the relative position of the Raisin River and the Put-in-Bay formations is decided by additional field work, preferably in Ohio and Michigan.

GENERAL CORRELATION OF CAYUGAN GROUP.

The Cayugan members present in Ontario are much more limited in numbers than are those of central New York state, where the Pittsford shale and the Vernon shale are present below the Camillus shale and higher formations occur above the Akron dolomite. Bassler correlates the Salina shales in a general way with the High Falls shale of eastern New York, the Longwood shale of New Jersey, and the McKenzie formation of Pennsylvania, Maryland, and Virginia. He correlates the Decatur limestone of west Tennessee with the lower Salina shales and the Leclaire (lower Clower) of Iowa and Minnesota and the lower part of the Waubakee limestone of eastern Wisconsin with the upper Salina shales (Camillus). In the upper Cayugan division Bassler includes, in ascending order: the Bloomsburg sandstone, the Wills Creek limestone, the Tonoloway limestone, and the Bossardville limestone, of Pennsylvania, Maryland, and Virginia, the Bertie and Akron being given as in part equivalent to the lower three divisions; the Sneedville limestone of Tennessee, given as in part equivalent to the Manlius of New York; the Kokoro limestone of Ohio west of Cincinnati, this being given as nearly the equivalent of the Bertie-Akron; and the Hillsboro sandstone and the Greenfield dolomite of southern Ohio and eastern Kentucky, given as representing the Bloomsburg sandstone and a part of the Manlius limestone respectively. The correlation of the Greenfield dolomite with the lower Manlius cannot stand if the writer's contention is correct, that the Greenfield dolomite is probably the equivalent of the lower part of the Bertie formation. Bassler correlates the Anamosa (upper Gower) of Iowa and Minnesota and the upper Waubakee limestone of eastern Wisconsin with the Bloomsburg sandstone or lowest part of the upper Cayugan. He also correlates the lower Cayugan with the upper part of the Stonehouse formation of Nova Scotia, and the whole Cayugan group with the Ludlow group of England, Scotland, and Wales. Twenhofel correlates the Stonehouse with the Ludlow of Europe, but with the Guelph of America.

There is reason to believe that formations of Cayugan age will be found in northern and western Canada when more detailed stratigraphy is done.

CAYUGAN HISTORY.

The diastrophism or regional movements which terminated Guelph sedimentation, caused the sea to withdraw from the greater part of North America, but left the following water bodies: a basin which included the present Great Lakes exclusive of lake Superior, and a portion of the Appalachian region as far south as southern Virginia; an ar i of the sea including part of Newfoundland, Anticosti island, id Gaspe peninsula; and probably areas in Florida, Alabama, Georgia, California, and southeastern Alaska which were connected with the open seas[1]. Unlike the other regions, the Great Lakes[1]-Appalachian basin was practically isolated from the sea during the deposition of the Camillus shale and particularly during the extreme desiccation which caused the precipitation of the salt beds. At this period the marginal uplifted sediments were eroded and deposited in the basin, and their salt content was added to the salt contained in any ocean water that i ight have been entrapped during uplift and warping. A return of i oro pluvial climatic conditions or an inflow of the sea from the Atlantic, brought about a period of less desiccation during which gypsum was deposited. Submergence with freer access to the Atlantic continued into Bertie time, during which, according to Miss O'Connell, rivers flowing from the north over flat-lying land formed deltas in which the remains of *Euryptcrida* were deposited along with nor i al i arine forms. An actual shrinkage of the basin area took place during the deposition of the upper Bass island, and the Bertie-Akron sediments, but contact with the ocean continued and the faunas consequently remained generally of marine character. The sediments, however, were principally derived from the surrounding emergent for i ations, as is shown by their being largely magnesian calcilutites. Under these restricted conditions of sedimentation, shale deposits varied greatly in character and thickness from place to place and the dolomitic deposits varied si i ilarly, as illustrated by the thin deposits in western New York and the thick deposits in Ohio and Michigan. Minor erosion intervals occurred i arked by slight disconformities as that seen in the Raisin River dolomite on East Sister island. Sand, probably eroded from the exposed St. Peter sandstone to the west, driven by the wind over barren deserts, finally formed arenaceous streaks in the upper dolomites prophetic of the close of Silurian time.

The end of Silurian sedimentation came in the Great Lakes region by a continued emergence of the land until all was above sea-level and the whole area was dry. Subaerial deposition of sand, rock flour, etc., went on, but this sedimentation is really epi-Silurian or transitional to the Devonian and as such sediments were largely modified by the encroaching Devonian seas, they are not described here.

[1]Bull. Geol. Soc., Am., vol. XX, 1910, pp. 421-606.

CHAPTER VI.

ECONOMIC GEOLOGY.

INTRODUCTION.

The economic products from the Silurian rocks of Ontario may be divided into two groups: the first including those products derived directly by quarrying or mining the rocks outcropping at the surface, such as building stone, road metal, gypsum, rock for flux, and cement, lime, brick, and tile manufacture; and the second those that have to be obtained by drilling such as salt, petroleum, and natural gas. One small zinc prospect, though not of economic importance, will be described under the first division.[1]

Although the formations vary much from place to place, it has been found quite possible to state in general terms for what purposes the rock of a formation can or can not be used. With products such as road metal and crushed stone for concrete and building stone, location is of great importance, and it is hoped that by the use of Maps 1714 and 1715 the best outcrops may be chosen for economic development.

The raw material contained in the Silurian formations, consists of magnesian limestone and dolomite of varying degrees of purity, sandstone of different colour and character, waterlime, shale, gypsum, rock salt, petroleum, and natural gas. Sphalerite, or zinc blende, galena, celestite, and some other minerals are scattered through the dolomites in small, unimportant quantities.

For detailed descriptions of the various industries and deposits, refer to the reports mentioned in the Bibliography on: building and ornamental stone by Parks; road metal by Reinecke; gypsum and salt by Cole; natural gas and petroleum by Clapp, Malcolm, and Knight.

BUILDING STONE.

MEDINA SANDSTONE.

An excellent quality of sandstone for building purposes has been obtained for many years from the Whirlpool member of the Medina-Cataract formation. This stone varies locally in colour from white or grey, to mottled and chocolate brown, and is best known as Credit Valley sandstone, because of the extensive quarrying carried on in the valley of Credit river. The buildings of the Legislative Assembly, Toronto, are built of Credit Valley brown sandstone.

The Whirlpool sandstone outcrops along the Niagara escarpment between Niagara river and a point about 8 miles south of Collingwood. The outcrops occur along the lower boundary of the Medina-Cataract formation, of which the Whirlpool sandstone is the lowest member in this region. The following general description is taken from Parks.[2] "The sandstone band of the Medina formation has an average thickness of about 12 foot only . . . . . most of the quarries exhibit irregular bedding, with lenses of one variety cutting diagonally across others. This irregularity has proved the greatest hindrance to the profitable extraction of the stone. Most of the beds are fine-grained and present roughly three

---

[1] This account of various raw materials used for economic purposes is supplementary to the information contained in the descriptions of the various geological formations and should be read with constant reference to the geological maps and sections.

[2] "Building and ornamental stones of Canada," vol. I, Dept. of Mines, Mines Branch, 1912, p. 139.

types; first, a brown to chocolate-coloured variety, quarried chiefly at the Forks of the Credit; second, a white or grey variety obtained at Orangeville and from Milton northward to the Forks of the Credit; also at Hamilton; and third, a mottled variety in which blebs and bands of white occur in the brown base. This latter kind is obtained more particularly from Merriton to Grimsby. The formation occurs along the face of the Niagara cuesta from near Merriton to Hamilton. Between Hamilton and Milton the cuesta is much broken up and the exposures are few, but from the latter place northward to Cataract, outcrops are numerous. Above Cataract, the stone is hidden very largely, but an important outcrop is known east of Orangeville and again near Shelburne. At no place is the exposure wide, as it appears only along the face of the 'mountain' and has scarcely any lateral extent. The widest and most accessible places occur where the sandstone forms the top of a shoulder, on the mountain side. Much of the quarrying has been done by removing a heavy overburden or by actually mining into the side of the hill. Unfortunately, those places where the stone is most easily obtained yield a poorer produce than where it is less accessible.

The more important areas for building stone are in the Credit Valley and to a lesser extent near Orangeville. Elsewhere, as in the Niagara and Milton areas, the stone is mostly used for rubble and rough coursing.

Dolomite and Magnesian Limestone. The Manitoulin dolomite is generally thin-bedded and argillaceous or clayey, and is consequently little used for other than rough coursing stone or rubble. In northeastern Owen Sound, about 25 feet of the Manitoulin beds are quarried for coursing stone; and a small amount of coursing stone has also been taken from small quarries (on lots 24 and 25, concession XI, Nottawasaga township, Simcoe county) about 8 miles south of Collingwood. In the vicinity of Manitowaning, Manitoulin island, the upper part of the Manitoulin dolomite is crystalline and probably could be worked locally for building stone. Much of the rock is thin-bedded, but at some localities it is very massive due to the presence of coral and bryozoan reefs; both of the extremes should be avoided in selecting a quarry site.

The Lockport (Niagara) dolomite, with its more or less local subdivisions, is generally suitable for building stone, and is accessible almost everywhere along the Niagara escarpment, and over large areas of Bruce peninsula and Manitoulin island. The DeCew waterlime is highly argillaceous and is consequently weak and unsuitable for building stone. The Gasport limestone of western New York is represented on the Canadian side by dolomitic limestone, highly crystalline and crinoidal in character. The Gasport loses its identity west and north of Dundas. The colour is generally dark blue, light blue, or grey, but some local areas are brown or pink. It is principally this stone which is quarried at St. Davids and Thorold where considerable amounts of dimension stone are produced.

The greater part of the Lockport formation below the Eramosa beds is grey or blue-grey on fresh fracture, weathering white, light, or creamy. Some of the lower beds are light yellowish when on fresh fracture. In a formation of such thickness much variation is to be expected. Pore spaces are numerous and large in some horizons and particularly in those that are fossiliferous, such as the " Pentamerus " beds occurring at the base of the formation in the Owen Sound district and northward.

Iron pyrites is common in the Pentamerus beds in some places and occu
also at other horizons. Good stone, however, is present in large amount
at most localities.

Extensive quarrying was formerly carried on near Beamsville in th
lower part of the Lockport formation. At Owen Sound, beds about ;
feet above the base are quarried. At Shelburne, beds high up in th
Lockport, just below the Eramosa division, are quarried. An examp
of this fine stone is to be seen in the Shelburne post-office (Plate XXXIV A

The upper beds of the Lockport (probably a little lower than th
Eramosa beds which have not been found on Manitoulin island), we
quarried extensively by the Lake Superior Corporation at Quarry poin
about 7 miles south of Meldrum bay, Manitoulin island, for the stone wo
of the Sault Ste. Marie canal, on the Canadian side. The stone weathe
grey, and has apparently been very satisfactory for the purpose for whi
it was used. Huge piles of waste at the quarry, however, show that mu
sorting of material was done. Many localities along the northern shor
of the western end of Manitoulin island and Cockburn island afford exc
lell opportunity of obtaining Lockport dolomite of good quality ac
near the water's edge.

The Eramosa dolomite is characteristically thin and even-bedde
At some localities these beds are almost as cleavable as slate, and towar
the top the partings are commonly bituminous. The colour is general
dark grey, banding of different shades being common. Near the city
Guelph, brown shades predominate.

The Eramosa beds outcrop below the base of the Guelph formation
many localities, but they and the Guelph are not known on Manitoul
island.

At a quarry west of Wiarton (lots 7 and 8, concession XIV, Amal
township, Bruce county) the upper Eramosa beds are worked. The thi
even-bedding, characteristic of this division of the Niagara, is here of gr
value in the production of stone of uniform thickness such as is requi
for monument bases and sills. The Eramosa beds are also quarried
Hamilton and Dundas, but at present the output is for road metal, concre
etc.

The Guelph dolomite varies much in character according to local
and horizon. The lower few feet are rather thick-bedded, but partake
the bituminous nature of the underlying Eramosa beds. The colour
consequently brown and a strong oily smell is common. These beds
not used for building stone. Beds higher up are commonly fine-grain
and dense, the bedding being thin and the colour light brown or gr
The beds near the top of the formation and those which appear to belong
the middle third are commonly light grey or cream-coloured, weathe
sandy. Pore spaces are numerous and are of various sizes, being parti
larly large in the more fossiliferous beds. The bedding in the upper divis
is generally thick to massive.

The lower, fine-grained Guelph beds are not used for building sto
although quarried at Fergus for crushed stone and lime. Beds a li
higher were formerly used at Fergus, Elora, and Guelph for local consti
tion, as were the upper fossiliferous beds at Hespeler and Galt. The st
from the higher beds is porous but cuts easily and gives a very plea
effect in the walls and trimmings of the buildings in which it has been u
Its wearing qualities are good. Large quantities of Guelph dolomite

available in the vicinity of the towns cited above, as well as at many other places within the Guelph area, especially on Bruce peninsula. As there is a considerable demand for crushed stone in the more thickly settled parts of the country, waste material from building stone quarries can be readily disposed of.

The Cayugan formations contain no beds that are very suitable for building. The Salina formation consists of shales and thin, impure dolomites in the lower part and brittle dolomites and waterlimes in the upper or Bertie division. The overlying Akron dolomite is thin-bedded and jointed. The Bass Island (Lower Monroe) formations of the Lake Huron and Detroit River regions are similar in character to the Bertie and Akron divisions with which they are correlated in this report.

ROAD METAL AND CRUSHED STONE.

The finer-grained, harder, Silurian dolomites are well suited for the construction of macadam roads subjected to medium traffic. The more porous, crystalline beds should be carefully avoided as the rock is too soft and crumbly to be useful for road metal.

The Manitoulin dolomite (Cataract) is quarried for coursing stone and road metal at Owen Sound and this material would doubtless also be useful for crushed stone for concrete work. L. Reinecke found this stone to be inferior in toughness, and to be suitable only for roads subjected to very light traffic. On the eastern part of Manitoulin island, the Manitoulin dolomite is very similar to the Lockport in character and would doubtless give good results for crushed stone and road metal.

The Clinton limestone thins out in Ontario and generally outcrops in the face of the Niagara escarpment in such a way as to be inaccessible for quarrying. The upper beds are tough and hard to work, particularly in the vicinity of Hamilton and Ancaster where they are known as the "Niagara Head" beds. The lower Clinton beds occur at the base of the bluff as far north as Kelso. Except for their inaccessibility, the Clinton beds would probably be useful for crushed stone and road metal.

The Lockport dolomite is extensively used for crushed stone, concrete and road metal. The largest quarry in this formation is that operated by the Canada Crushed Stone Corporation, Limited, at Dundas. The lower 32 feet of strata here are the Eramosa beds and the upper 25 are Guelph. This stone has been favourably reported on for road metal by Reinecke[1]. According to his report it is suitable for roads subjected to light traffic and is one of the very best of the limestone-dolomite group in the country. Other quarries are operated in the Lockport at St. Davids, Thorold, Hamilton, Ancaster, Waterdown, Orangeville, and elsewhere. According to Reinecke, tests made on the stone at Ancaster show it to be very similar to that at Dundas. Stone of the general characters of that already tested could be obtained at many places along the Niagara escarpment and at other localities over the extent of the Lockport outcrops.

The Eramosa beds at the top of the Lockport dolomite are cleavable, siliceous and bituminous in part, and as shown above are found suitable for road metal at Ancaster, Dundas, and elsewhere. Outcrops of these beds are to be found east of Guelph, in the quarries at Ancaster and Dundas, at the quarries for dimension stone west of Wiarton, at Halfway Rock, on

Road Mat. Res., Mem. 85, 1915, p. 255.

the north shore of Bruce peninsula, and on Flowerpot and other islands in Georgian bay. These beds are generally capped by tough Guelph strata with which they could be quarried to good advantage, as is done at Ancaster and Dundas.

The Guelph dolomite varies much in character, but where fine-grained is very resistant and furnishes splendid material for crushed stone for concrete, artificial stone, and road metal. The quarries operated at Galt and Fergus furnish good stone for such purposes. Reinecke reports that the stone from the Ciow quarry, Fergus, is suitable for light traffic only. It has, however, worn well on local roads. The ubber 2 or 3 feet at Ancaster and the over 25 feet of strata quarried at Dundas are Guelph. A large quarry is operated by the Canadian Pacific railway for ballast, about one mile southwest of Shaw station, and quarries are operated at Rhymal and elsewhere. Many opportunities occur for opening quarries in the Guelph rock over the extent of its outcrops, but where road metal is required, only the hardest and densest stone should be selected; a selection can generally be made by examining the surface exposures.

The Bertie-Akron dolomites furnish good road metal in the vicinity of Byng near Dunnville. Similar rock occurs west of Cayuga, and in the bend of the Saugeen river about $2\frac{1}{2}$ miles north of Walkerton (see page 89). The outcrops of rock at Innerkip and on the islands west of Pelee island, belonging to the Bass Island series, would be serviceable for crushed stone and road metal.

CEMENT.

Magnesian limestones carrying considerable quantities of argillaceous (clayey) matter, and known as waterlimes, were formerly burned in kilns to a natural rock cement which possessed the characters of a low grade Portland cement. With the standardizing and increased manufacture of high grade cement, the natural rock cement industry has died out.

DECEW WATERLIME BEDS.

At one time considerable amounts of the DeCew beds were mined at St. Davids, Thorold, and DeCew falls, for the manufacture of natural rock cement. Quarries were also worked in these beds at Mount Albion and probably elsewhere. Six feet of cement rock was worked at St. Davids, and 8 feet at Thorold and DeCew falls. At St. Davids and Thorold the rock was mined on the room and pillar system, the entrances being levels run from nearby quarries. The cement manufactured at Thorold was used in the building of the Welland canal and Victoria bridge. For analysis of the DeCew beds see page 112. The DeCew beds are available for mining at various places along the Niagara escarpment between Hamilton and Niagara river.

BERTIE WATERLIME.

About 6 feet of Bertie waterlime was formerly quarried at North Buffalo and burnt to natural rock cement. Beds of similar character occur at quarries north of Ridgeway (lot 1, concession X, Bertie township) and in the quarries at and near Byng across the Grand river from Dunnville. Should a demand arise for such rock, it may be obtained at the places mentioned and also in the vicinity west and south of Bridgeburg, Ontario.

PORTLAND CEMENT.

No limestones low enough in magnesia for the manufacture of Portland cement are found in the Silurian system of southwestern Ontario. Such limestones should be sought in the areas underlain by Ordovician (e.g. Trenton) and Devonian (e.g. Onondaga) formations.

DOLOMITE FOR FLUX.

Dolomite or high magnesian limestone is required in the sulphite process of paper-manufacture from pulpwood and also for flux in the basic open-hearth process of steel manufacture. Because dolomite may be obtained more cheaply for some blast furnaces, for example at Hamilton, it is used for flux instead of limestone, which is otherwise preferred. Dolomites from Rhymad and Dundas are used in the Hamilton blast furnaces. About 30 per cent of the total production, including the entire upper bed of the rock quarried at Dundas by the Canada Crushed Stone Company, is sold for flux for blast furnaces. From the analyses (pages 110 to 116), the characters of the Silurian dolomites may be studied. The distribution of the various formations is given in the chapter on stratigraphy.

MANUFACTURE OF CHEMICALS.

High grade limestone such as is required for chemical manufacture is not found in the Silurian formations. This is to be found in the Onondaga and Anderdon limestones at Amherstburg and Beachville'.

Dolomite is suitable for the manufacture of carbon dioxide, and the purer forms such as are found in the Niagara and Guelph formations are easily available for such purposes.

DOLOMITE FOR LIME MANUFACTURE.

Magnesian limestone and dolomite burn to quicklime which makes strong mortar, although it is somewhat slow in setting. For the manufacture of hydrated lime, the presence of considerable quantities of magnesium carbonate in the rock is essential.

Formerly all the Silurian dolomites were burnt to some extent for quicklime. The lime-kilns were largely operated by farmers to supply the local demand, wood being used as fuel. The remains of old kilns near rock outcrops, in nearly all parts of the country, testify to the source of the lime used in local building enterprises.

The Manitoulin dolomite was burnt for lime on many parts of Manitoulin island, and especially near Manitowaning and Gore Bay. The Lockport dolomite was formerly burnt at many localities along the Niagara escarpment, and is still burnt at Ancaster, Kelso, Limehouse, Rockwood, Owen Sound, and elsewhere.

Guelph dolomite is burnt to quicklime at Puslinch, Hespeler, Galt, Elora, Fergus, and in the vicinity of Guelph, where it is also manufactured into hydrated lime by the Standard White Lime Company.

The Bertie-Akron dolomite was formerly burnt to quicklime at Springvale, Innerkip, $2\frac{1}{4}$ miles north of Walkerton along the Saugeen river, and 2 miles south of Pinkerton.

Geol. Surv., Can., Mem. 34, 1915, pp. 271, 274, 279.

Great quantities of raw material are available for the manufacture of hydrated and ordinary lime. It is obvious that in choosing a quarry site, cherty dolomite and dolomite full of clayey impurities should be avoided. For analyses of the various rocks see pages 110 to 116.

SHALE FOR BRICK AND TILE MANUFACTURE.

The Cabot Head shale as known on the mainland of Ontario is too sandy or else contains too many calcareous beds to be useful in the manufacture of brick or tile. A considerable thickness of nearly uniform, red, plastic shale is found, however, in the Cabot Head shale of Manitoulin island. A sample of this shale collected by the author on the road allowance on lot 16, concession IX, Billings township, was submitted to J. Keele, chief of the Ceramic division, Mines Branch, Department of Mines, who reported on it as follows:

" This shale is very plastic, smooth, and stiff, when tempered with water. Owing to its stiffness it is hard to work, and would be improved in this respect by the addition of sand.

The material dries without cracking, with shrinkages within working limits, and burns to a hard, dense, red body at low temperatures.

This clay appears to be suitable for the manufacture of building brick, hollow ware blocks, and field drain tile. It may be possible to use it for roofing tile also."

The shale from which the sample was taken occurred in a 50-foot section exposed along the road for about one-eighth of a mile. It is situated about 2½ miles from West bay, Indian village. A smaller exposure occurs on lot 28, concession IX, Allan township, about 3 miles southeast of the town of Gore Bay. The clay is exposed in the road on both sides of a small hill, which is capped with an impure dolomite. Large quantities of clay could doubtless be obtained below the surface soil on the flanks of the hill.

Other less favourably located outcrops of this shale occur on Manitoulin island, one of the best being on lot 27 on the road between concessions d V, Bidwell township, and 1¼ miles west of lake Manitou.

ZINC BLENDE.

Small particles of galena and sphalerite, or zinc blende, are widely scattered through the Niagara and Guelph dolomites. Particles of galena may be found in the Lockport dolomite at the top of the falls at Grimsby, at Rockwood, and east of Guelph. Zinc blende may be seen at Rockwood and elsewhere.

The only notable amount of zinc blende reported, however, was discovered by Dr. Wolverton of London, on lot 30, concession II, Albemarle East township, Bruce peninsula, about 4 miles northwest of Wiarton. Under the direction of Mr. G. B. Bourne considerable prospecting has been done since , an excavation having been made measuring approximately 30 feet in length, 10 feet wide, and 25 to 30 feet in greatest depth (not 100 feet by 30 feet by 30 feet, as reported in the Summary Report, 1912, page 281). The rock is massive, very scraggy, and full of pore spaces and cavities, and when fresh is a light blue colour. The large amount of pore spaces and cavities is principally due to the solution of fossils which are large and numerous, some of the cephalopods being over 1 foot long and 6 inches in diameter. The zinc blende occupies these cavities, the mineral in some cases partly replacing fossils. The action of surface waters has in some cases caused a concentration of pieces of zinc blende at the bottom of small pockets, into which the blende has fallen on being released from the rock by solution of the dolomite to which it was attached. About 110 pounds of loose blende is said to have been obtained in such a pocket, where it was mixed with soil and pieces of rock. The ore continues down to a depth of 20 feet, below which no ore has been found.

A shipment of several hundred pounds of this ore was made previous to 1912, the lot being said to assay 69·76 per cent zinc. When last visited (September 1916) about two cords of hand-picked ore was piled up alongside the prospect.

The zinc blende occurs in dolomite of lower Guelph age, closely related in fauna to the Racine beds of Wisconsin. Not far below the zinc deposits is the top of the Eramosa, arenaceous dolomite forming the upper division of the Lockport formation. These beds, because of their fine-grained, impervious nature have doubtless prevented downward movement of the surface waters, which on this account moved entirely in a lateral direction, dissolving such minerals as zinc blende under favourable conditions for solution and precipitating them where numerous pore spaces and cavities existed owing to the solution of fossils. The accumulation in local pockets is doubtless very recent, and due to surface solution.

GYPSUM.

Extensive gypsum deposits occur in the Salina[1] formation of western New York and of the adjoining Niagara peninsula of Ontario as far west as the vicinity of Paris. The gypsum occurs near the top of the Camillus shale and just below the Bertie waterlime (see the description of the core drill record at Buffalo). Some gypsum occurs at the top of the shale throughout Ontario peninsula, as we may judge from evidence of well borings. In the west and north, however, gypsum occurs in small amount and salt occurs in large quantities lower down in the Salina. For further discussion of the Salina formation and the origin of the gypsum deposits, see Chapter V, page 84.

The workable deposits of gypsum, as at present known, occur along Grand river between the old gypsum mines, 4 miles southeast of Cayuga, and old workings about 1 mile northeast of Paris. Grand River practically follows the outcrop of the gypsum, which being quite soluble has been easily eroded away. Cole[2] has mapped the workable gypsum deposits in such a way as to separate the above region into two areas, a western area included within a radius of 2 miles of Paris and not definitely determined towards Brantford; and an eastern area of oblong shape extending from a point about 4 miles east of Brantford to the bend of Grand river 4 miles west of Dunnville, the greatest width being about 6 miles in the vicinity of York. These areas contain the workable deposits at present known, the only actual mining being carried on in the vicinity of Caledonia and York. Well records, however, indicate gypsum throughout Niagara peninsula, although the exact thickness is not indicated in most logs of wells kept by

[1] Formerly known as the Onondaga formation: Onondaga is now properly used for a Devonian formation = "Corniferous."
[2] "Gypsum in Canada," Dept. of Mines, Mines Branch, 1913, p. 76.

104

the gas co1panies. In New York state, exploration is being done at more than 100 feet in depth and it is the cost of production compared with the market price which will deter1ine the li1it in depth at which the Ontario gypsu1 may be worked. As the average southerly dip per 1ile of the geological for1ation is about 30 feet, it will be seen that $3\frac{1}{2}$ miles south of the outcrop, a bed of gypsu1 will be approxi1ately 100 feet lower than at the outcrop. Differences in the elevation of the surface of the land 1ust, of course, be taken into account in esti1ating the depth of the gypsum below the surface. The gypsu1 is without doubt lenticular, but the horizon is extensive and there see1s no reason to suppose that deposits as good as those already known may not yet be discovered in Niagara peninsula, at, however, greater depths than those now being worked. Under present conditions, the cost of working such deposits would doubtless be prohibitive.

For detailed discussion of the gypsu1 industry of Grand valley, see "Gypsu1 in Canada."[1]

On January 1, 1917, the two co1panies for1erly operating in Grand River valley, viz., the Alabastine Company, Li1ited, of Paris, and the Crown Gypsu1 Company, of Lythmore, a1alga1ated under the na1e of the Ontario Gypsu1 Co1pany, Li1ited. The Alabastine Company 1anufactured wall tints, known as alabastine, and wall plaster, as well as ground gypsu1 for retarder in the 1anufacture of portland ce1ent and for fertilizer or land plaster. The Crown Gypsu1 Company 1anufactured wall plaster. The operations of the two co1panies are now carried on by the Ontario Gypsum Company.

The principal 1ine for1erly worked by the Alabastine Company is on lots 10 and 11, concession 1, west of the Ha1ilton road, Seneca township, Haldimand county. About 15 acres have been 1ined, operations being carried on in part on three different levels. The roo1 and pillar 1ethod of 1ining is followed, the roo1s on the second level being 20 feet square and the pillars 12 feet; the roo1s on the third level are 18 feet and the pillars 14 feet. Underground haulage is done by means of horses and the gypsu1 is finally raised to the surface along an incline by stea1 power. Self-du1ping cars carrying the gypsu1 in steel half cylinders are used. Little water is encountered, the mine being so dry and co1fortable that the visiting 1e1bers of the Twelfth International Geological Congress were served dinner by the Alabastine Company in August, 1913, in the 1ain station of the 1ine on the second level.

The product of this mine is green gypsu1 which is used in the 1anufacture of land plaster and as a regulator for the setting of portland ce1ent. Large reserves are held north of the present workings.[2]

The Carson 1ine "located about three miles south of Caledonia village in Oneida township" supplied the Alabastine Company with pure white gypsu1 for its finer products. The bed of gypsu1 worked[3] is about 4 feet thick, and 1ining is carried on as at Caledonia. "Overlying the bed of white gypsu1 are layers of dolo1itic shales and li1estones, in all from 4 to 6 feet in thickness, covered by 40 to 50 feet of post-Glacial drift. The roof here is not in as safe a condition as at Caledonia, where the grey beds are from 9 to 12 feet in thickness, overlying which are 40 to 50 feet of dolo1itic li1estone and shales covered by the post-Glacial drift."

[1] Ibid, pages 57 to 77.
[2] Ont. Bureau of Mines, 1915, vol. XXIV, pt 1, p. 150.
[3] Ont. Bureau of Mines, 1915, vol. XXIV, pt. 1, p. 150.

The mine formerly operated by the Crown Gypsum Company is situated near the village of York on the south side of Grand river, on lots 58 and 59, Oneida township, Haldimand county. This is known as the Martindale mine.

The following is taken from Cole's description, page 67.

"The property on which this Company's working mine is situated is one of the first places in this district from which gypsum was mined. With the exception of about twelve years, this gypsum bed has been operated continuously since about 1866.

The beds are covered with about 70 feet of drift, shale, and limestone. The one that is being worked is reached by an incline tunnel 500 feet in length with a bearing of north 73 degrees east. This bed consists of very high grade, white gypsum suitable for the best grades of plaster of paris, and breaks clean to both floor and back so that very little waste has to be sorted out underground. Hand power is employed for drilling the waste rock in the mine. The bed has an average thickness of about 4 feet and as only the gypsum is broken, the workings are consequently very low thus making moving around underground very difficult.

The method of mining the gypsum in this mine is very similar to the room and pillar method employed in coal mines....."

For descriptions of the abandoned mines and prospects see Cole's report.

The output of gypsum for Ontario reached a maximum in 1914[1], the production being: gypsum mined, 89,157 tons; gypsum shipped, crushed, or ground, 43,183 tons; ground and calcined gypsum used in gypsum products, 33,705 tons; the manufactured products being valued at $162,375 and the total production $221,175.

The total output in 1917, crushed, ground, and calcined,[2] was 48,656 tons valued at $128,828. This is a falling off of 44,744 tons since the maximum production of 1914, which was 93,400 tons of gypsum and gypsum products.

SALT.

The total Canadian production of salt is obtained from the Camillus shale of the Salina formation of the Ontario peninsula. As shown on Cole's map,[3] the productive salt deposits are approximately included in the area lying between lake Huron, St. Clair river, and lake St. Clair on the west, and a line from the shore of lake Huron at Inverhuron south through Brussels haif-way between Seaforth and Dublin, to the east of London and to lake Erie shore half-way between Dutton and Port Stanley. The southern boundary extends from near the Elgin-Kent boundary on lake Erie shore through Thamesville, and to the shore of lake St. Clair a few miles north of the mouth of Thames river. A small area in the vicinity of Windsor is also salt-producing. The salt occurs in several beds which are more or less lenticular, as is illustrated by the occurrence of 125 feet of salt (with some shale beds) in a well bored on the south side of Thames river about 2 miles west of Prairie siding, although no salt was reported in the wells drilled all around this locality. Another similar occurrence is that of 400 feet of salt penetrated in a well

[1] Ont. Bureau of Mines, 1915, p. 42.
[2] Ont. Bureau of Mines, Bull. 33, 1917, p. 3.
[3] "Salt deposits of Canada and the salt industry." Dept. of Mines, Mines Branch, 1915, opposite p. 20.

in the Wheatley oil field. The most extensive gypsum deposits do not occur in the salt area, but considerable gypsum is found in the Camillus shale overlying the salt. Some gypsum is interbedded with the salt at Goderich and gypsum occurs below the salt at Courtright. However, owing to the small amount of gypsum present and to its lower solubility than salt, little trouble is experienced in the manufacture of salt from Ontario brines.

Several beds of salt may be penetrated by a well; for example six occur at Goderich, the total thickness of salt being 126 feet. At Windsor, four beds are reported with a total thickness of 427 feet.

Salt is obtained by evaporating the brine pumped up from the wells. The water for using this brine is allowed to flow into the wells, usually from a water-bearing stratum high above the salt. Evaporation is accomplished by steam heat or by a vacuum process.

Besides the general production of table, dairy, and coarse salt for meat packing, etc., caustic soda and bleaching powder are manufactured by the Canadian Salt Company at Sandwich.

Salt was first discovered by Mr. Platt at Goderich in 1866, while drilling for oil. At present, salt is manufactured at Goderich, Kincardine, Wingham, Clinton, Exeter, Parkhill, Elarton, Sarnia, Courtright, Windsor, and Sandwich. The total production[1] for 1917 was 138,528 tons, valued at $1,095,866. This is the maximum production to date. In 1916, 305,900 pounds, valued at $2,223, were exported and 151,208 tons, valued at $694,835, were imported, including 109,493 tons, valued at $523,725, imported from Great Britain for the use of coast fisheries.[2]

Analyses of Brine from Rock Salt, Southwestern Ontario[1].
Hypothetical combination: 1,000 parts by weight contain:

	I	II	III	IV	V	VI	VII	VIII
NaCl	244·860	253·105	266·415	258·770	263·921	260·812	245·111	256·891
CaCl$_2$	1·265	1·066	1·895	1·484	0·477	0·111	4·964	1·007
MgCl$_2$	0·966	0·534	1·884	1·017	2·040	0·297	0·503	0·467
CaSO$_4$	728	4·037	2·757	3·730	3·939	4·476	2·405	3·971

I. Western Canada Flour Mill's well, Goderich, Ont.
II. American Chemical Co.'s well, Goderich, Ont.
III. Stapleton Salt Works, Clinton, Ont.
IV. Ontario Peoples Salt and Soda Co., Kincardine, Ont.
V. Sparling Co., Wingham, Ont.
VI. Western Salt Co., Mooretown, Ont.
VII. Dominion Salt Co., Sarnia, old well.
VIII. Dominion Salt Co., Sarnia, new well.

Analyses of Brine from the Platt Well at Goderich[4].

—	Parts per 1,000.	In 100 parts of solid residue.
NaCl	259·000	99·018
CaCl$_2$	0·432	0·165
MgCl$_2$	0·254	0·097
CaSO$_4$	1·882	0·720
	261·568	100·000

[1]Ont. Bureau of Mines, Bull. 33, 1917. p. 3.
[2]Mines Branch, Dept. of Mines, Sum. Rept., 1916, p. 174.
[3]"Salt deposits of Canada," table III, p. 48.
[4]Geol. Surv., Can., 1866, p 269.

As no analyses of the l ines fro i the Windsor wells are at hand the following analysis is given to represent this region.

Artificial Brine Direct from the Well of the Delray Salt Com y, Delray, Michigan, near Detroit[1].

KCl	trace
NaCl	290·6
CaCl₂	trace
MgCl₂	3·6
CaSO₄	3·1
MgBr₂	trace

Analyses of Brines from New York State, for Comparison[2].

	I	II
MgCl₂	0·155	0·049
CaCl₂	0·129	0·134
CaSO₄	0·599	0·349
NaCl	16·921	23·295

I. From the Onondaga natural brine.
II. From artificial brine from the solution of the rock salt.

Analyses of Rock Salt from Attrill Well, Goderich[3].

	I	II
NaCl	99·687	91·24
CaCl₂	0·032	0·57
MgCl₂	0·095	0·05
CaSO₄	0·090	2·81
Insoluble	0·017	5·33
Moisture	0·079	
	100·000	100·000

I. From the white translucent portion of the second bed in the Attrill well core.
II. From the less pure salt from the same bed.

The following co i parison is by Hunt.[4]

	Impurities.
	Per cent.
Cheshire salt	2·67
Cordova, Spain, salt	1·45
Turks Island salt	2·34
Saginaw salt	2·00
Syracuse solar salt	1·15
Syracuse boiled salt	1·50
Goderich coarse salt (analysis made in 1871)	1·09
Goderich medium salt	1·28
Goderich fine salt	1·62

[1]Cook, C. W., Mich. Geol. and Biol. Surv., Pub. No. 15, geol. ser. 12, p. 89.
[2]Newlands, D. H., N.Y. State Mus., Bull. 174, p. 68.
[3]Geol. Surv., Can., 1866, p. 260.
[4]Geol. Surv., Can., vol. V, pt. II, p. 164 s.

For further comparison, the following analyses of rock salt are added

Analyses of Rock Salt from New York State.

	I	II
Moisture . . .	trace.	trace.
Insoluble . . .	0·7430	0·0584
CaSO₄	0·4838	0·0793
CaCl₂ .	0·0180	0·0358
MgCl₂ .	0·0546	0·0888
NaCl........	98·7000	99·7410

I. An average sample of salt mined in New York state.
II. " Perfectly white salt."

The superior quality of the Ontario salt is clearly illustrated by comparing the above analyses.

PETROLEUM.

Of the petroleum obtained from southwestern Ontario, only a small part comes from Silurian formations, the most of it being found in the Onondaga (Corniferous) limestone of the Devonian system. Some oil, however, is obtained from the Whirlpool sandstone in Onondaga township, Brant county, and from the top of the Guelph and the lower Salina of the Tilbury field and the Guelph of the Wheatley field.

The oil production from the Whirlpool sandstone is obtained[2] from Onondaga township, Brant county, about 7 miles east of Brantford along Fairchild creek, and from an area approximately 1½ miles long by 1 mile wide. The Whirlpool sandstone is here 20 feet thick and lies 500 feet below the surface. This field was opened in 1910, reached its maximum production of 13,501 barrels in 1911, and produced 1,617 barrels in 1916.

As interpreted from drill records, the upper Guelph and lowest Salina strata yield oil in the Tilbury field, which is situated in the eastern part of East Tilbury and the western part of Raleigh townships, Kent county. Including the gas field which extends to lake Erie and includes the eastern point of Romney township, the field is 9 miles in length from north to south and 8 miles in greatest width which is along the shore of lake Erie. In the vicinity of Stewart, East Tilbury, gas is reported from a depth of 1,363 feet and oil from 1,393, 1,418, and 1,430 feet.

Oil is also obtained from the Guelph formation in the Wheatley field at a depth of 1,290 and 1,300 feet. This field includes a small area east of Wheatley in the southwestern part of Romney township.

The production of oil for Tilbury and Romney, which was 106,993 barrels in 1906, reached a maximum of 411,588 barrels in 1907 and then rapidly fell off, being 16,297 barrels for 1916 (as reported for Tilbury presumably including the Wheatley field)[3] which was a gain of 3,553 barrels over 1915, the lowest output since the opening of the field.

Oil in Ontario is generally found to occur at the top of domes and anticlines, where it is driven by the greater specific gravity of salt water. In the case of formations that contain no water this rule does not hold

[1]Merrill, Frederick G. H., Bull. N.Y. State Mus., No. 11, 1893, p. 37.
[2]Geol. Surv., Can., Mem. 81, 1915, p. 61.
[3]Ont. Bureau of Mines, vol. XXVI, 1917, p. 52.

but the Silurian formations containing oil generally contain salt water also. The present tendency is to explore the Guelph and Salina formations, where not already tested, below the domes and anticlines discovered in the development of the shallower oil fields (Onondaga or "Corniferous") (see structure, pages 11 to 14).

GAS.

Most of the gas found in Ontario is from formations of Silurian age. The Medina-Cataract sandstones and shales, the Clinton limestone, the Guelph dolomite, and the Salina shale all bear locally important quantities of gas.

The Medina sandstones are generally gas-bearing. Quoting Malcolm[1] "The greater proportion of the gas of the Selkirk, Port Dover, and Delhi fields is derived from the red Medina sandstone (Grimsby member), although small quantities are found also in the white Medina sandstone (Whirlpool member), and in the Clinton formation. The White Medina sandstone is the productive horizon for the gas of the Bertie-Humberstone, Wainfleet, Ottercliffe, Bayham, Onondaga, and Brantford fields. . . . Some of these fields derive a portion of their gas also from the red Medina sandstone and the Clinton formation.' For the depths at which the top of the Medina is reached in the various fields, see Malcolm, page 29, and cross-sections and logs in this report.

The Clinton formation, according to Malcolm, supplies gas "in the Onondaga, Caledonia, and Cayuga fields, and is the most productive horizon at Port Colborne, west of the canal." For depths, see Malcolm, page 5., and logs given in this report.

Malcolm states "the Guelph is the gas-bearing formation of Gosfield, Mersea, and Wheatley. . ." At Gosfield, the top of the formation is about 1,000 feet below the surface and at Mersea about 1,100 feet. For further information as to depths, see Malcolm, page 34.

"The Salina (Onondaga) is the gas-bearing formation of the Kent gas field. . ." (Malcolm, page 36, see page 37 for depths of top of formation.)

For the chemical composition of the natural gases of Ontario, see 23rd Annual Report of the Ontario Bureau of Mines, 1914, pp. 247-259.

It is now well recognized that the occurrence of natural gas and oil in the Silurian formations of Ontario is principally dependent upon rock porosity and structure such as anticlines, domes, monoclines, etc., and consequently developing companies are paying more and more attention to lenses of porous strata and to structure as worked out from well logs. The keeping of careful logs of the formations penetrated in drilling, and the depth at which water is struck, etc., is of great importance for the future development of oil and gas in Ontario (for structure see page 11).

According to the Bureau of Mines[2] the total gas production for Ontario in 1916 was 17,953,000,000 cubic feet, valued at $2,404,499, as compared with 15,211,500,000 cubic feet in 1915. Of this amount, Welland-Haldimand produced 3,769,500,000, Kent 13,752,500,000, Elgin 351,-900,000, and Lambton 55,200,000 cubic feet. The amount for 1917[3] was 20,025,700,000 cubic feet, valued at $3,182,154.

[1] Ibid, page 28.
[2] Ont. Bureau of Mines, vol XXVI, 1917, p. 46.
[3] Ont. Bureau of Mines, Bull. 33, p. 3.

ROCK AN ..YSES.

MEDINA-CATARACT FORMATION.

Manitoulin dolomite, from Mitchell's Mills, lot 8, concession XI Collingwood township, Grey county. Analyses by Mines Branch, Dept of Mines, Ottawa.

$CaCO_3$	66·00
$MgCO_3$	20·54
Fe_2O_3	1·05
Fe_2O_3	0·06
Al_2O_3	0·73
Insoluble	10·70
Total	99·08

Analyses of Manitoulin Dolomite on Manitoulin Island[1]

	I	II	III	IV	V
Insoluble matter			3·86		
Silica	4·64	5·68		7·64	2·76
Ferric oxide	1·12	0·81	0·72	0·94	0·5
Alumina	1·80	2·85	0·32	2·44	0·5
Lime	28·61	28·08	29·00	26·59	24·5
Magnesia	19·60	19·10	20·48	18·82	20·3
Carbon dioxide	43·79	43·00	45·27	41·57	45·67
Loss		0·16	0·22	0·25	
Sulphur trioxide		0·30	0·12	0·51	0·18
Alkalis			0·66		
Total	99·47	99·98	100·65	98·56	99·66

    I. Gore Bay, sample of 12 feet of uppermost part of face of cliff, northwest of
      Fair grounds.
    II. Top of cliff across bay, east of Gore Bay.
    III. Gore Bay, one-quarter mile west of northwest corner of Fair gro...ds, cliff
      6 feet.
    IV. Porter's quarry, just east of Fair grounds, Gore bay.
    V. Landing at lake Manitou.

Calcareous sandstone bed thought to represent the southern extension of the Manitoulin dolomite, as taken by the author in the Niagara gorge above Lewiston. Analysis by Mines Branch, Dept. of Mines, Ottawa.

$CaCO_3$	38·39
$MgCO_3$	23·95
$FeCO_3$	3·14
Fe_2O_3	0·07
Al_2O_3	3·42
Insoluble matter	32·38
Total	101·35

[1] Ont. Bureau of Mines, 1904, pt. II, p. 75. The formation is identified by the writer from his knowledge of the localities described.

A sample taken from a 6-foot bed of ferruginous limestone in the upper part of the Cabot Head shale at Limehouse. Analysis by Mines Branch, Dept. of Mines, Ottawa.

$CaCO_3$.	28·84
$MgCO_3$	21·59
$FeCO_3$.	2·05
Fe_2O_3.....	20·72
Al_2O_3	2·48
Insoluble	22·18
Total	97·80

CLINTON FORMATION.

Clinton, basal, sandy, glauconitic layer about 2 or 3 inches thick, at Dundas. Analysis by Mines Branch, Dept. of Mines, Ottawa.

SiO_2............	22·42
Al_2O_3.................	3·46
FeO.	2·72
Fe_2O_3.....	2·97
CaO.....	19·80
MgO.......	13·92
H_2O.	3·50
CO_2	not determined.
Total.....	68·79

Reynales (Wolcott) dolomite at Hamilton. Analysis by Mines Branch, Dept. of Mines, Ottawa.

$CaCO_3$........	50·85
$MgCO_3$.....	35·28
$FeCO_3$.......	2·11
Fe_2O_3.....	0·07
Al_2O_3	0·50
Insoluble........	10·90
Total	99·71

Irondequoit dolomite, at Hamilton. Analysis by Mines Branch, Dept. of Mines, Ottawa.

$CaCO_3$ . .	53·69
$MgCO_3$....	39·19
$FeCO_3$.	2·38
Fe_2O_3.....	trace
Al_2O_3...........	0·39
Insoluble.........	4·24
Total.........	99·89

NIAGARA GROUP.

Analyses of Rochester Shales.[1]

	I	II
SiO_2.......	23·00	22·52
Al_2O_3........	12·65	8·12
FeO.....	1·77	1·13
Fe_2O_3......	0·46	1·01
CaO.....	20·81	21·83
MgO......	7·77	10·14
H_2O.....	7·20	4·00
Total...........	72·66	68·75

I. Rochester shale, 3 feet above base at Hamilton.
II. Rochester shale, upper 4 feet, at Hamilton.

[1]Analyses by Mines Branch, Dept. of Mines, Ottawa.

Analyses of DeCew Waterlime Beds, Lockport Dolomite.[1]

	I	II
SiO₂	21·70	14·48
Al₂O₃	5·88	5·10
FeO	0·90	0·67
Fe₂O₃	1·28	1·26
CaO	23·62	20·29
MgO	12·98	15·96
H₂O....	1·00	5·60
CO₂	not determined.	
Total	67·36	63·36

I. About middle of beds at St. Davids.
II. Redhill creek, west of Mount Albion.

Analyses of DeCew Waterlime, from Thorold Quarries, Used in Piers of Victoria Bridge.[2]

	I	II
Calcium carbonate.....	56·28	47·02
Magnesium carbonate	20·07	30·32
Alumina and ferric oxide.......	2·20	2·71
Insoluble argillaceous residue..	20·90	19·64
Water and loss by ignition	0·31	0·29
Total....	99·76 ·	99·94

I. John Brown's quarry.
II. Alex. Manning's quarry.

Rock at base of Lockport formation at Limehouse formerly burnt for hydraulic lime; probably DeCew waterlime[3].

Carbonate of lime..		48·07
" magnesia		39·63
" iron .... : . . .		0·69
Sulphate of lime		0·10
Alumina... . .	0·21	
Silica, soluble..... .	0·37	
Insoluble matter, consisting of:		
Silica	7·60	11·60
Alumina.	2·07	
Ferric oxide.......	0·40	
Lime......	0·05	11·02
Magnesia..	0·19	
Potassa.......	0·53	
Soda..	0·18	
Total..		100·09

[1]Analyses by Mines Branch, Dept. of Mines, Ottawa.
[2]Ont. Bureau of Mines, 1904, pt. II, p. 15.
[3]Geol. Surv., Can., vol. VIII, 1895, pt. R, p. 16.

Analyses of Lockport Dolomite [1]

	I	II	III
CaCO₃ ...	55·30	54·28	53·48
MgCO₃ ...	42·28	39·00	42·55
Fe₂O₃ .	0·82	1·66	0·24
FeO₂ .	0·14	0·57	0·07
Al₂O₃	0·15	0·29	0·12
Insoluble ..	1·24	3·00	3·50
Total ...	99·93	98·80	99·96

I. 3-foot bed above 6-foot bed of Reynales (Wolcott) dolomite, at Kelso.
II. Beds about 23 feet above DeCew beds, at St. Davids.
III. Government quarry at Providence Bay, Manitoulin island.

Analyses of Lockport Dolomite from Manitoulin Island. [2]

	I	II
Insoluble ...		
Silica ...	1·43	1·42
Ferric oxide...	0·71	0·99
Alumina,...	0·59	...
Lime ...	30·05	30·06
Magnesia ...	21·19[3]	20·46
Carbon dioxide	46·00	47·00
Total	99·97	99·93

I. Top of hill at Meldrum Bay village.
II. Fossil hill, near Manitowaning.

Analyses of Lockport dolomite from Ryan and Haney's Quarry, near Meldrum Bay, Manitoulin Island. [1]

	I	II
Silica	0·40	0·56
Ferric oxide	0·50	0·41
Alumina	trace	0·20
Lime	30·84	30·50
Magnesia	21·11	21·55
Carbon dioxide ...	47·40	47·40
Loss ..	0·18	
Sulphur trioxide	0·09	0·23
Alkalis	0·15	0·05
Total	100·67	100·90

[1] Analyses by Mines Branch, Dept. of Mines, Ottawa.
[2] Prel. Bureau of Mines, 1904, pt. II, p. 75.
[3] Given as 11·39, which is evidently incorrect. The figures were corrected from the total as printed

114

Lockport dolo ite fro Cockburn island formerly used for manufacturing sulphite pulp at Sault Ste. Marie[1].

Insoluble	4·5
Iron and alumina	0·5
Calcium carbonate	52·0
Magnesium carbonate	41·0
Undetermined	2·0
	100·0

Brownish grey, crystalline, Lockport dolo ite, Gri sby.[2]

Carbonate of lime	68·92
Carbonate of magnesia	29·48
Carbonate of iron	1·10
Insoluble	0·50
	100·00

Lockport dolo ite used in the anufacture of li e at Christic's siding near Kelso, west half of lot 3, concession VI, Nassagaweya township, Halton county[3].

Carbonate of lime	54·12
" magnesia	45·45
" iron	0·58
Sulphate of lime	0·17
Alumina trace	0·30
Insoluble matter 0·30	
	100·62

Analysis of rock fro Ancaster quarries, presu ably fro the Eramosa beds[4].

Moisture	0·23
Insoluble matter	1·60
Carbonate of lime	53·30
Carbonate of magnesia	43·13

Average of eight sa ples of rock fro lower quarry of Canada Crushed Stone Co pany at Dundas, representing about 15 feet of upper Eramosa beds. Analysed by Crowell and Murray, ining engineers, Cleveland, Ohio. Analysis furnished by Mr. Doolittle, manager of Canada Crushed Stone Co pany.

Silica	1·66
Iron oxide	0·54
Alumina	0·37
$CaCO_3$	54·42
$MgCO_3$	43·02
Sulphur	0·034
Phosphorus	0·015
	100·059

GUELPH DOLOMITE.

Analysis of average of eight sa ples of upper 16 to 18 feet of quarry at Dundas, being the lower part of the Guelph for ation. Analyses by

[1] Ont. Bureau of Mines, 1904, pt. II, p. 13.
[2] Geol. Surv., Can., 1876-77, pp. 486-7.
[3] Geol. Surv., Can., vol. VIII, 1895, pt. R, p. 17.
[4] Ont. Bureau of Mines, 1903, p. 142.

Crowell and Murray, mining engineers, Cleveland, Ohio, and supplied by Mr. Doolittle, manager of Canada Crushed Stone Co i pany.

Silica	0·41
Iron oxide	0·43
Alumina	0·59
CaCO₃	55·20
MgCO₃	43·60
Sulphur	0·016
Phosphorus	0·005
	100·251

Analyses of Guelph Dolomite, Wellington County.

	I	II
Carbonate of lime	54·25	53·97
" magnesia	45·17	45·37
" iron	0·22	0·16
Sulphate of lime	0·34	0·68
Alumina	trace	trace
Insoluble matter	0·08	0·03
Total	100·06	100·21
(Hygroscopic water after drying at 100° C.)	0·05	0·02

I. Light grey, fine crystalline, massive dolomite from Wellington quarry, south half of lot 29, gore of Puslinch township, Wellington county.

II. Light, cream-yellow, yellowish-brown weathering, very fine crystalline, compact, dolomite, from the priest's quarry on Speed river, Guelph township, Wellington county.

SALINA FORMATION.

Camillus Shale.

Lowest beds exposed in Caledonia gypsum mine. Analysis by Mines Branch, Dept. of Mines, Ottawa.

SiO₂	12·34
Al₂O₃	3·93
FeO	1·94
Fe₂O₃	0·41
CaO	24·20
MgO	16·17
H₂O	3·00
	61·99

CO₂ not determined.

Shale accompanying gypsum at Paris.[2]

Water	0·75
Silica	52·02
Alumina	8·03
Ferric oxide	3·80
Calcium carbonate	9·90
Magnesium carbonate	2·34
Sulphur	1·00

[1] Geol. Surv., Can., vol. VIII, pt. R, 1895, p. 17.
[2] Ont. Bureau of Mines, 1904, pt. II, p. 34.

Thin li**estones, which are interbedded with shale below gypsu horizon at Paris.

Water	0·33
Insoluble	3·32
Calcium oxide	27·77
Magnesium oxide	15·15
Carbonic acid	33·42
Sulphur	0·60

BERTIE-AKRON FORMATION.

Bertie Waterlime.[1]

Rock cut by Welland canal on north end of lot 28, concession II, Hu**berstone township, Welland county[2].

First Analysis.

CaO	44·68
MgO	36·27
Silica, c... ..d insoluble silicates	16·14
Fe_2O_3 and Al_2O_3	2·46
	99·55

Second Analysis.

SiO_2	12·32
Fe_2O_3, Al_2O_3	4·94
CaO	25·02
MgO	16·80
CO_2	39·13
Moisture	0·06
	98·27

Waterlime at Springvale[3].

Lime	31·58
Magnesia	17·79
Alumina	3·29
Ferric oxide	1·89
Silica	3·69
Water	0·15
Carbonic acid (ignition loss)	44·73

Waterlime, probably Bertie. Bluish, lower rock in quarry on lot 40, concession IV, North Cayuga township, Haldimand county[4].

Water	0·55
Silica	4·14
Alumina	26·60
Ferric oxide	1·56
Calcium oxide	20·09
Magnesium oxide	14·51

Dolomite, probably Akron, occurring below Oriskany sandstone at Michigan Central cut 2 miles east of Hagersville[5].

Water	0·35
Silica	3·44
Alumina	2·34
Ferric oxide	1·86
Calcium oxide	26·61
Magnesium oxide	17·47
Ignition loss	44·96

[1] Some Akron dolomite may be included.
[2] Ont. Bureau of Mines, 1902, p. 31.
[3] Ont. Bureau of Mines, 1904, pt. II, p. 118.
[4] Ont. Bureau of Mines, 1903, p. 116.
[5] Ont. Bureau of Mines, 1903, p. 144.

117

CHAPTE⟨ ⟩I.

PALÆONTOLOGY: DESCRIPTION OF NEW SPECIES AND PALÆONTOLOGICAL NOTES.

MEDINA-CATARACT FORMATION.

Lingula cuneata Hall.

The original description[1] is as follows: "Shell cuneifor⟨m⟩, very acute at the beaks; ⟨m⟩argins nearly rectilinear, and converging unifor⟨m⟩ly fro⟨m⟩ near the base to the beak; base slightly curved; the valves convex on the upper half but flattened or co⟨m⟩pressed towards the base; surface longitudinally striated.

This ⟨s⟩pecies is readily distinguished fro⟨m⟩ any of the other species by its acutely cuneata for⟨m⟩ .... but the shell is more or less exfoliated and the striæ partially or wholly destroyed. The shell is translucent and usua!⟨l⟩y of a light colour, particularly in the light grey sandstone......"

Fro⟨m⟩ an exa⟨m⟩ination of a ⟨q⟩uantity of ⟨m⟩aterial fro⟨m⟩ the Whit⟨e⟩ ore ⟨q⟩uarries, Lockport, N.Y., it is found necessary to ⟨m⟩odify the above description as follows: on practically every slab exa⟨m⟩ined there are valves of two distinct shapes, the acute form as described above and another for⟨m⟩ with ⟨m⟩ore obtuse beak. The latter for⟨m⟩ evidently represents the dorsal valve. A few ventral valves have subparallel sides, the beaks being acute and the general shape cuneifor⟨m⟩. A very shallow ⟨m⟩edian septu⟨m⟩ is present in the ventral valves. This extends fro⟨m⟩ about 2⟨m⟩⟨m⟩. fro⟨m⟩ the beak to about 5⟨m⟩⟨m⟩. fro⟨m⟩ the frônt where it ends in "central scars". The septu⟨m⟩ is bounded by "concrete laterals" 1⟨m⟩⟨m⟩. or less apart. The septu⟨m⟩ and acco⟨m⟩par⟨y⟩ing parts are well illustrated in "Palæontology of New York"[2]. This figure also illustrates the less cuneate ventral valves already ⟨m⟩entioned. The shell on such parts as are well preserved shows fine radiating striæ, about four to five being spaced in 1⟨m⟩⟨m⟩. The exfoliated surfaces show fibrous structure, rese⟨m⟩bling striæ.

The typical shells at Lockport occurring in red sandstone, average: ventral valve 18 to 19⟨m⟩⟨m⟩. long, 9 to 12⟨m⟩⟨m⟩. wide at front; dorsal 15 to 18⟨m⟩⟨m⟩. long, with width as in ventral. In grey sandstone lenses, the length is generally less and the shape less cuneate. At a slightly higher horizon, occurring in highly ferruginous sandstone along with bryozoa and ostrocoda, lingulas are very nu⟨m⟩erous but are only 7 to 10⟨m⟩⟨m⟩. long. In other characters they are indistinguishable fro⟨m⟩ the well-developed specimens found not 100 yards away.

At Niagara river, in a 3-foot bed of arenaceous li⟨m⟩estone 23 feet above the Whirlpool sandstone, fragmentary lingulas occur; these are typical *L. cuneata*. Twenty-two feet higher up in red sandstone this species of lingula is again abundant, but averages only about 12⟨m⟩⟨m⟩. in length. The lingulas of the latter horizon are abundant also near Thorold. At Grimsby and Ha⟨m⟩ilton, lingulas occur in grey sandstone near the top of the Medina section, the beds being older than the true grey band or Thorold quartzite. The lingulas of these localities are of about the sa⟨m⟩e size as those of the upper horizon at Niagara river. but are generally less cunei-

[1]Palæontology of New York, vol. II, 1852, p. 8.
[2]Vol VIII, pt. I, pl. I, fig. 11.

for 1. In all other respects they are si 1ilar to the typical *L. cuneata* and there are a 1ong the 1 so 1 e decidedly cuneate for 1s. However. Schuchert identifies these as *Lingula clintoni* Vanuxem, for 1 erly known as *L. oblonga*.

Dalmanella eugeniensis new species.

(Plate VII, figures 1 to 6.)

Shell wider than long, sub-oval 1 ore or less, planoconvex, hinge-line shorter than greatest width of shell. Striæ fine, angular, recurved fro 1 u 1 bo to lateral 1 argin, increasing by intercalation on dorsal valve and do 1 inantly by bifurcation on ventral valve. One or two well-1 arked growth lines present near anterior 1 argin of shell. Length of adult speci 1 ens fro 1 12 to 18 1 1.; width 14 to 22 mm.; thickness 4·6 1 1.

Ventral valve 1 oderately convex, flattened toward anterior 1 argin and concave between beak and cardinal-lateral angle; u 1 bo 1 erging into general curvature of shell; beak s 1 all, acute, scarcely incurved: cardinal area 1 oderately high, tapering rapidly to cardinal-lateral angles; deltidial plates narrow; pedicle opening wide; hinge teeth large and strong; 1 uscular area short, divided medially by low, rounded elevation.

Dorsal valve nearly flat, with narrow 1 edian depression which becomes shallow and ill-defined at front. Beak acute, very short; cardinal area very narrow, cardinal process s 1 all trilobed; 1 uscular area about one-third length of shell.

This species closely rese 1 bles *Dalmanella elegantula* (Dal 1 an), from which it may be distinguished by being wider than long, having a low broad u 1 bo, cardinal-lateral angles of very little 1 ore than 90 degrees, s 1 aller cardinal process, shorter crura, and being generally of larger size.

Occurrence. Cataract for 1 ation at Eugenia, Ont., in lower 10 feet of Cabot Head grey shale; at Lavender falls near Dunedin, Ont., in shale partings at top of Manitoulin dolo 1 ite; at Ha 1 ilton, Ont., in red shaly beds of Gri 1 sby sandstone; near Owen Sound, Ont., along Pottowattomi river in Cabot Head shale just above Manitoulin dolo 1 ite.

Dalmanella eugeniensis var *palæoelegantula* Willia 1 s n. var.

(Plate VII, figures 7 and 8.)

S 1 all, se 1 i-circular, plano-convex; ventral valve moderately deep, u 1 bo 1 oderately high and narrow, 1 ore rounded in younger specimens; cardinal-lateral angles rounded and hinge-line about one-half the width of the shell; dorsal valve nearly flat or gently convex near the beak: a narrow, 1 edian groove fro 1 beak to front, anteriorly at the botto 1 of a shallow sinus; length of type speci 1 en 1 c 1., width 1 cm., thickness at u 1 bo 3·5 1 1.

This variety differs fro 1 *D. eugeniensis* in being nearly as long as broad, in having a higher, narrower u 1 bo, and, in the type speci 1 en, in having a narrow, 1 edian groove on the dorsal valve. Younger specimens have 1 ore convex dorsal valves and lack the median groove.

This variety approaches the characters of *D. elegantula* (Dal 1 an), and hence its variety na 1 e: it differs fro 1 the later Silurian species, however, in being proportionately wider, in having a lower umbo, and in having a smaller cardinal process and shorter crura.

Occurrence. In Cabot Head shale at Eugenia (collected by J. Stansfield and H. V. Ellsworth 1911), and in Manitoulin dolorite at Glen William.

MEASUREMENTS OF VIRGIANA MAYVILLENSIS AND PENTAMERUS OBLONGUS.

Virgiana mayvillensis Savage, *Cataract Formation.*

	Fitz-william Island.	Tamarack cove.	South bay.	West bay.	Kaga-wong lake.	Helen bay.	Cockburn island.
Specimens examined	4	1	4	2	3	6	5
Smallest	2·1	....	2·5	1·9	1·9	2·2	2·7 cm. long.
Largest	3·0	....	3·9	3·0	2·3	3·5	3·5 "
Average	2·5	2·1	3·1	2·4	2·2	2·6	3·1 "
Average ventral septum	0·5	1·7	0·9	0·9	1·1	1·1	0·9 "
¹Average ratio	2·8,2·5	3·5,4·0	2·4,2·2	4·8,4·4	3·7,2·0	3·2 ",1·3·1,2·8

¹The average ratio is an expression of the quotients obtained by dividing the average lengths and widths of the ventral valves of the specimens successively by the average depth.

Pentamerus Oblongus in Reynales (Clinton) *Dolomite.*

	Rochester.	Thorold.	DeCew falls.	Jordan.	Grimsby.	Kelso.
Specimens examined	5	4	1	2	3	2
Smallest	3·1	2·0		2·0	1·7	5·0 cm. long.
Largest	5·7	6·0		4·0	5·8	6·2 "
Average	4·5	4·2	3·0	3·0	3·7	5·6 "
Average ventral septum	1·9	2·5		0·2	1·9	2·5 "
Average ratio	3,2·3	com-pressed.	1·9, 1·7	1·6, 5·2	5·3, 4·7	2·8, 2·8

Pentamerus oblongus occurs from 1·5 to 2 feet above the base of the Reynales dolomite.

Pentamerus Oblongus in Lockport Dolomite.

	Inglis mills.	Owen Sound	Wiarton.	Barrow bay.	Cabot head, 25 feet up.	Cabot head, near base.	Fossil hill.	Lake Manitou.	Cockburn island.
Specimen examined	4	3	4	8	1	3	2	7	3
Smallest	2·5	5·0	3·7	2·6	....	5·7	3·4	2·2	3·3 cm. long.
Largest	5·5	6·0	5·2	6·0	....	6±	5 ±	3·7	8·0 "
Average	4·3	5·5	4·8	4·5	3·5	5·8	4·2±	2·9	5·6 "
Average ventral septum	2·2	3·6	2·6	3·1	2·5	1·8	2·8	1·9	3·0 "
Average ratio	3·9	4·6	2·8	3	4	2·9	6·0	2·6	4·3
	3·6	4·3	2·2	2·5	3	2	5	2·3	3·3

Rhynchotreta cabotensis, new species.

(Plate VII, figures 11 a, b, c.)

Shell small, compressed, sub-triangular, not broader than long, valves sub-equally convex. Plications 12 to 14, prominent, rounded, extending from beak to front; separated by interspaces which, at the front, are equal to or greater than the width of the plications. Growth lines prominent near the front only. Length and breadth between 6 and 9 mm., depth 2 to 5 mm., angle between cardinal slopes 70 degrees to 100 degrees or more in some cases.

Pedicle valve moderately convex; beak small, acute, gently incurved foramen slit extending from the foramen to the contract with the beak of the brachial valve; umbo moderately well defined; median sinus containing two plications confined to the front half of the shell.

Brachial valve slightly more convex than pedicle; beak small; a shallow sinus from the beak to the middle of the shell containing three plications which rise as a low fold at the front of the shell.

The two small figured specimens measure respectively 7 mm. long, 6·5 mm. wide, 2·5 mm. deep; and 6 mm. long, 7 mm. wide, and 2·5 mm. deep. The brachial valve of the larger figured specimen, which is of about average size, is 8·5 mm. long, 9 mm. wide, and 2 mm. deep.

R. cabotensis differs from *R. intermedia* Savage in being smaller and less transverse. From *R. lepida* Savage, with which it most closely compares, it differs in the presence of a fold and sinus, and in greater proportional breadth.

Occurrence. In Dyer Bay dolomite and the overlying Cabot Head shale at the "Clay bank" along the shore about 2 miles west of Cabot head.

Atrypa parksi new species.

(Plate VII, figures 19 a, b, c.)

Length and breadth sub-equal. Beak of pedicle valve closely incurved and covering the beak of the brachial valve. Both valves convex, the pedicle with a well-marked sinus in front, not appearing in the umbonal region, and the brachial with a fold, well-marked in front, where the margin of the shell is turned up in some specimens, and only faintly indicated or forming a shallow sinus near the beak. Shell widest posteriorly; cardinal slope varying and extremities rounded. Surface marked by fine rounded plications (5 to 2 mm.) which increase generally by bifurcation and less commonly by implantation and by growth lines which are obscure except near the front of the shell. The specimen chosen as the type measures 24 mm. wide, 18 mm. high, and 10 mm. deep at the umbo. Other specimens are from 13 to 25 mm. in width. Well preserved, small specimens at Credit Forks approach the form of *A. marginalis*, having decided cardino-lateral angles, small, prominent, pedicle beaks, and straight frontal margin at the fold.

This species is most closely related to *Atrypa marginalis* Dalman from which it differs in having considerably finer plications, rounded cardinal extremities, less clearly defined fold and sinus, and in being of larger average size.

The species is na i ed after Professor W. A. Parks, of Toronto University, who recognized it as a new species, but left it for the author to describe in this i e i oir.

Occurrence. In the Manitoulin dolo i ite at Stoney Creek, Ha i ilton, Dundas, Credit Forks, Owen Sound, Cabot head, and Church hill, Manitowaning, and West bay, Manitoulin island.

Whitfieldella cataractensis new species,

(Plate VII, figures 16 to 18.)

Elongate oval, i oderately convex, rounded, cardinal-lateral angles for i ing decided "shoulders"; narrow, shallow sinuses on anterior three-fourths of both valves; greatest convexity in posterior third; pedicale valve i ore convex than brachial; average length $1 \cdot 7$ cm., width $1 \cdot 3$ c i., depth $1 \cdot 0$ c i. A large fairly well preserved speci i en i easures: length $2 \cdot 3$ cm., width $1 \cdot 7$ c i., and thickness, allowing for crushing, about $1 \cdot 2$ cm. Another speci i en doubtfully referred to this species, has no sinus in the brachial? valve, which i easures: length 2 c i., width $1 \cdot 6$ c i., depth (one valve) $0 \cdot 5$ c i. Surfaces of the shells are i arked by growth lines only.

The nu i ber of speci i ens found is over forty, but i any of these are mere frag i ents, and all are crushed or worn. It is probable that the sinuses have been accentuated by crushing. A i ong the frag i ents are remains of so i e s i all speci i ens and also of so i e large ones that i ave widths of fro i 2 to $2 \cdot 6$ c i.

W. cataractensis is i ost nearly allied to *W. quadrangularis* Foerste, fro i the Brassfield of Ada i s county, Ohio. Fro i this species it differs in being s i aller on the average, less rotund, and in having i ore clearly defined and narrower sinuses.

Occurrence. Manitoulin dolo i ite at Hamilton, Glen Willia i, and Lavender falls; in the arenaceous Manitoulin beds at Niagara river; in li i estone lenses in the Cabot Head shales at Stoney Creek, near Owen Sound, and at Credit Forks.

Nuculites (Cleidophorus) cf ferrugineum Foerste.

(Plate VIII, figure 4 (s i all shells).)

S i all, elliptical, beak about one-third the distance fro i the anterior end; a notch and short, narrow sinus in front of the beak. The larger of the two figured speci i ens is $1 \cdot 3$ c i. long and 7 i i. deep; this is an average speci i en. The s i aller speci i en is 1 c i. long and $0 \cdot 6$ c i. deep.

The above specimens are larger and more elongate than the single specimen figured by Foerste; in other respects there appears to be no difference.

Occurrence. Gri i sby sandstone at Kelso, and at "Sydenha i road", West Fla i borough tonwship, Wentworth county, where it was collected by R. Bell.

57237—9

Pterinea brisa Hall?

Species represented by left valve. Hinge line about 1·6 cm. in length. Greatest radial distance from beaks 1·5 cm.

Valve compressed, oblique. Anterior wing unknown; posterior wing distinctly separated from body and terminated in an angle of more than 90 degrees. The most distinctive character of this species is the presence of five, close set (1 mm. apart), wavy lines of growth on the body of the shell. The undulations are about 1 mm. in amplitude. On the posterior wing the lines of growth are curved, with convex side forward.

Another specimen, probably belonging to this species, is decidedly convex, measures 1 cm. on the hinge line, and 1 cm. in length. It has fine, radiating striæ about three in 1 mm. and has three decided growth lines, 1·5 mm. apart, on the anterior half of the shell.

In outline this species resembles *P. elegans*, Savage.

Locality. Grimsby sandstone, at Kelso, Ont.

Pterinea cf *undata* (Hall).

(Plate VIII, figure 2.)

Left valve observed: oblique, moderately convex, length of hinge line about 2·5 cm.; greatest radial length from beak 3 cm. Anterior wing rounded, separated from body of shell by a shallow sinus situated about one-third the distance from the anterior to the posterior extremity. Posterior wing not distinctly separated from the body of the shell, and not extending beyond length of shell. Hinge line making obtuse angles with outline of shell at both extremities. Umbo extending slightly beyond hinge line. Surface marked by growth lines and striations, two to three in 1 mm., the striations being seen only near the centre of the shell in this specimen.

Two small specimens measure, respectively: hinge line 1·4 cm. length 1·4 cm.; and hinge line 1·2 cm., length 1·4 cm. The broader shows faint traces of radiating striæ, and both show growth lines, the broader approaching *P. undata* in characters.

This species resembles *P. undata* Hall closely in outline, but differs from it in surface markings which may, however, be exfoliation characters only. It corresponds closely with the fragment figured by Hall. "Palæontology of New York", volume II, 1852, Plate 4, bis, figure II d and e. These figures have been removed from *Modiolopsis primigenus* where Hall placed them, by Bassler[1] who has not identified them. Figure IIc (Hall) compares with this species in outline, but is a left valve, and being convex is probably not a *Pterinea*. Hall's specimens were from the Medina sandstone at Medina.

Occurrence. Grimsby sandstone, Niagara gorge south of Lewiston N.Y.

[1] U. S. Nat. Mus. Bull. 92, vol. II, p. 819.

Modiolopsis Kelsoensis new species.

(Plate VIII, figure 3.)

Shells oval, and moderately convex; beak about one-third the distance from the front of the shell; a shallow but well-marked sinus from the centre of the beak downward and somewhat backward to the base of the shell; marked growth lines over the entire surface; hinge area not seen; length of type 3·5 cm., depth 1·5 cm., thickness, including both valves, about 1·4 cm.

Occurrence. In red Grimsby sandstone at Kelso, and on lot 11, concession 1, West Flamborough, Wentworth county. The latter locality is represented by a specimen collected by R. Bell and formerly labelled *Grammysia Canadensis* Billings (see figures).

Lophospira pulchra new species.

(Plate VII, figures 22a, b.)

Shell conic, slightly wider than high. Height 17 mm., width 15 to 16 mm., apical angle about 85 degrees, volutions 4, slit band prominent; a well-marked keel present on all volutions above slit band, about half-way between slit band and suture. Suture impressed, younger volutions overlapping on ohler about half-way to slit band, which is centrally located.

Surfaces between slit band, keel, and suture, and slit band and suture, concave. Lower surface of last volution subventricose. Fine striations, directed somewhat backward, cross the surfaces from the suture downwards to slit band, being undisturbed by the keel. From the slit band to the suture below they are directed forwards. Aperture unknown.

Occurrence. At Cataract, in flint nodules in lower 6 feet of Manitoulin dolomite. Collector, J. Stansfield, 1912. Only one specimen is known to the writer.[1]

Observation. This species is very similar to *L. bispiralis* (Hall), of the Guelph formation, from which it differs, however, in being somewhat lower and consequently having a larger apical angle and in having the keel above the slit band persistent on all the volutions including the outer one.

Pleurotomaria species.

One specimen was found by the writers' assistant, H. V. Ellsworth, at the top of the Whirlpool sandstone, in a small gully near the cement tunnel over the New York Central tracks, south from Lewiston, New York. This specimen is fragmentary and about two-thirds the size of Hall's figured specimens of *P. littorea*. It is also proportionately lower. In other respects it corresponds with this species.

A number of casts of small gastropods occur in the upper beds of Whirlpool sandstone at Glen William, Ont. They appear to be all of one species and correspond closely to *H. subulata* as described and illustrated. They, however, show faint, but distinct "peripheral angulation and band" and so differ from *H. subulata* as so far known (Plate VIII, figure 1).

[1] A small undetermined species of Lophospira, about 4 mm. in height and greatest diameter, occurs in the same chert nodule.

124

ROCHESTER FORMATION.

Strophonella deeewensis new species.

(Plate XIII, figure 1.)

Brachial valve, semi-oval, longer than broad, moderately convex, broadest and most convex near the centre; lateral areas in front of the hinge-line concave, area in front of beak flattened and containing a short narrow sinus; striæ fine, rounded, and curved backward near the hinge line, number about 100, increasing by implantation; hinge-line less than width of shell; cardinal-lateral angles rounded. Length 3·4 cm., width 3·2 cm. Pedicle valve not seen.

This species differs from *S. patenta* in having greater length than width and rounded, cardinal-lateral angles.

Occurrence. Rochester shale, 8 to 12 feet below the top at Det's falls near Power Glen, Lincoln county, Ont.

LOCKPORT FORMATION.

Stricklandinia manitouensis new species.

(Plate XX, figures 2, 3.)

Longer than wide, very convex; valves nearly equal; greatest convexity in posterior portion of shell; a faint sinus in the pedicle valve and a faint fold on the brachial valve; simple, rounded, radiating striæ, about five in 5 mm. at the front of the shell; median septum about 3 to 4 mm in length. The beaks are worn in all three specimens in hand, but the beaks of both valves are evidently short, the pedicle closely incurved over the brachial. The following measurements are only approximate owing to the worn condition of the specimens: No. 1, length 2·5 cm., width 2·1 cm., thickness 1·8 cm.; No. 2, length restored about 3 cm., width 3·2 cm., thickness 1·5 cm.; No. 3, length 3 cm., width 2·6 cm., thickness 1·9 cm.

This species is thicker and narrower than *S. deformis* Meek and Worthen and lacks the growth lines of that species. It compares most closely with *S. norwoodi* Foerste, from the figures of which it differs in having the fold and sinus less marked, and in having more nearly marked striations, and a tendency to greater rotundity.

Formation and Locality. Lockport dolomite, Fossil Hill horizon, plateau east of Sandfield, Manitoulin island.

GUELPH FORMATION.

Orthoceras brucensis, new species.

(Plate XXVI, figures 1 a, b.)

Straight, gently tapering, about 2½ mm. in the space of one diameter surface marked by oblique annulations, about four in the width of a segment; cross-section circular, siphuncle about half-way between cen-

and circumfer ce; diameter 1·2 to 2·7 cm. Camaræ deep, in the shape of short cones ter inated by hemispheres, thickness unusual, 1·5 to 1·7 mm. The centres of the ca aræ weather out so as to leave only the walls preserved, the result suggesting double camaræ.

This species is very si ilar to *O. unionense* Worthen, but differs in having deep, thick camaræ.

Occurrence. Guelph dolo ite on Bruce peninsula at Hay bay, 3 iles west of Tobermory, at Dorcas bay and Baptist harbour near Cape Hurd, and in Lockport dolomite about 15 feet above the base at Limehouse, Ontario.

APPENDIX I.

NEW SPECIES OF BRACHIOPOD.

(By A. F. Foerste.)

Rhynchotreta Williamsi new species.

(Plate VII, figures 15 a, b.)

This species is closely related to *Rhynchotreta thebesensis* Foerste, from which it differs chiefly in the absence of any tendency toward the elevation of the two median plications of the brachial valve, and a corresponding depression for the median plication of the pedicle valve. Apparently, the plications are much less distinct and the beak of the pedicle valve is more strongly incurved, but it is not certain whether these differences are not due in part to the state of preservation of the single specimen here described as *Rhynchotreta williamsi*. In this specimen, the plications are distinct only near the anterior margin and are nearly obsolete posteriorly. The specimen was obtained, however, in a dolomitic rock and may be one of those casts partaking more of the interior characters of the species than of its exterior surface features. Eleven or twelve plications may be distinctly recognized along the anterior margin of the shell, in addition to which two or three plications are faintly indicated on each of the flattened sides.

Rhynchotreta williamsi was found in the Manitoulin dolomite, about 2 miles southeast of Manitowaning, on Manitoulin island, where the dolomite forms a conspicuous hill, south of the road to South bay. The species is named in honour of M. Y. Williams, with whom the writer had the pleasure of visiting the locality in question.

Rhynchotreta thebesensis was found in the Edgewood limestone division of the Medina, at Thebes, Illinois.

APPENDIX II.

NEW SPECIES OF CRINOID.

(By Frank Springer.)

Eucheirocrinus Ontario new species.

An imperfect specimen, the only one found in rocks of this formation, the geological position being older than that of *E. chrysalis* (Hall), and *E. radiculus* (Ringueberg), both from the Rochester shales. Enough is preserved, however, to show the essential characters on which the species is based, which are:

The left anterior ray (="dorsal arm") bifurcates on the third plate above the radial, and on every third brachial above that, by equal branches; thus it has three primibrachs. The left posterior ray has only two primibrachs, bifurcating unequally on the second, and on every second plate above that. The two segments of the left anterior radial are in contact, the inferior segment being a wide plate for its entire length, with parallel sides. These characters are shown in Figure 6.

The species differs from *E. chrysalis* and *E. radiculus*, from Lockport, in the much smaller number of primibrachs in the dorsal arm, those species having seven to nine or more; and in the greater width and more rectangular form of the inferior segment of the radial. In the former respect it resembles a species from the Wenlock formation at Dudley, England, which has only one primibrach (perhaps "*Cheirocrinus*" *sagittalis* Salter, catalogue name). In the latter it is somewhat comparable to *E. barrandei* Walcott, from the Trenton of New York, and recalls the very wide corresponding plate in *Cremacrinus*.

Occurrence. Manitoulin bed, Cataract formation, Stoney Creek, Ontario.

FIGURE 6. *Eucheirocrinus Ontario* n. sp.
Diagrammatic view. Magnified
2 diameters. G.S.C.
Mus. No. 4730.

APPENDIX III.

NEW SPECIES OF CORAL.

(By George H. Chadwick.)

Family Columnariidæ.

Genus *Palæophyllum* Billings.

Palæophyllum (Cyathophylloides?) williamsi sp. nov.

(Plate V, figure 2.)

Corallum composite, sub-fasciculate; corallites cylindrical; the new corallites interspersed among the older, from 1½ mm. up to rarely 6 mm. in diameter; the mature corallites averaging about 4 mm. Gemmation appears to be lateral and fairly prolific. Outer wall strong; exterior surface slightly undulate and with septal striæ; a strongly wrinkled epitheca like that of *Cyathaxonia cynodon* is preserved at a few points on the holotype. Sepia usually about eighteen to twenty-four in adult corallites, not reaching the centre, rather strong and regularly arranged but occasionally very slightly twisting or curving; in addition, short, interstitial septa alternate with the long ones but are in many cases rudimentary. The central area is well-like and has a diameter varying from nearly equal to the length of the long septa to less than half that amount. The tabulæ vary from strongly convex to flat or even concave in the centre with the periphery curved down, and are spaced from 1 to 2 mm. apart; they do not interrupt the longer septa. No dissepiments nor fossula. Type specimen collected by M. Y. Williams from the Manitoulin dolomite east of Manitowaning bay, Manitoulin island.

Observations. This appears to be the species most likely to be identified as "*Diphyphyllum multicaule*" in the Cataract formation. The habit of growth is exactly that of *Synaptophyllum* except for the absence of the connecting processes, but the internal structure is wholly different, being that of *Columnaria*. The composite rather than compound mode of growth would appear, however, to be of generic value.

This species differs from *Palæophyllum stokesi* (Edwards and Haime) of the Richmond, by its smaller corallites and more widely spaced tabulæ also in the entire absence of the connecting processes mentioned by Edwards and Haime and figured by Whiteaves. In its mode of growth and arrangement of tabulæ it is still less like the Richmond genotype, *P. rugosum* Billings (as figured by Lambe), which is clearly distinct specifically from *P. stokesi*. Much more closely allied is *Palæophyllum thomi* (Hall) of Texas. This last is referred by Walcott to *Cyathophylloides* Dybowski whose typical species supposedly include *C. rhenanus* Frech and *C. disjunctus* (Whiteaves) of the Devonian. Should this reference prove appropriate, perhaps in a subgeneric sense as used by Whiteaves, it will carry with it the other two thick-walled species which do not underg polygonal compression, *P. williamsi* and *P. stokesi*. The corallites in *P. williamsi* are more variable in size than shown for *P. thomi* and have not the peculiar structure in transverse section represented in Hall's figures, but comparison with actual specimens may leave few real distinctions.

Palæophyllum umbellicrescens sp. nov.

Corallum subcompound, usually small, subastraciform. Corallites closely set, subparallel or radiating, tapering cylindrical or becoming polygonal toward the base by mutual compression, varying from $\frac{3}{4}$ to 3 mm. in diameter, the mature polyps being seldom under $2\frac{1}{2}$ mm. Epitheca well marked, obscuring the septal grooves where not weathered away. Tabulæ convex, numerous and complete, spaced $\frac{1}{2}$ to $\frac{3}{4}$ mm. apart, not interrupting the septa. Septa short, the alternate ones reaching about half-way to the centre, in some cases a little more; the longer septa, about twelve in number in mature corallites, regularly arranged, with as many intermediate ones.

When a corallite reaches maturity (3 mm.) the inner ends of the longer septa become united in a continuous ring, like an inner wall, thus producing a whorl of marginal chambers each with a single septum (the short intermediate one). These calices immediately assume a circular form, remaining in contact laterally, and new septa rapidly appear as the growing corallite enlarges upward. By crowding the circular arrangement is soon lost, but the umbels spring again with considerable regularity, simultaneously from each corallite, after from 15 to 20 mm. of upward growth.

The type specimens (six in number) were collected by M. Y. Williams from "The Rock", and a seventh specimen was obtained by M. Y. Williams from the "Devil's Needle", both localities being near Manitowaning, Manitoulin island. Formation, Manitoulin dolomite.

Observations. The more closely aggregated corallites, closer tabulæ and thin wall, ally this species with the genotype, *P. rugosum* Billings of the Richmond, from which and from *P. calcinum* (Nicholson) its small size and fewer, shorter septa at once distinguish it. The peculiar multiplication by septal generation in marginal whorls as in *Romingeria* has not previously been remarked in *Columnaria* or *Palæophyllum*, and may prove of subgeneric value.

PLATE II.

57237—11

PLATE IV.

Medina-Cataract Fossils.

(Figures natural size.)

FIGURE 1. *Favosites cristatus* Edwards and Haime, St. Edmund dolomite at Cabot head. G.S.C., Mus. No. 4506. (Page 38.)

FIGURE 2. *Helopora fragilis* Hall. Grey dolomite lens in Cabot Head shale, Stoney Creek, ont. G.S.C., Mus. No. 4507. (Page 35.)

Plate IV.

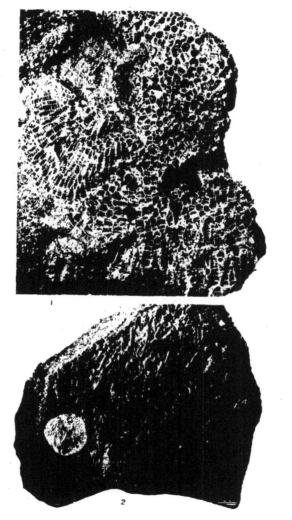

1

2

Plate V.

Medina-Cataract Fossils.

(Figures natural size.)

FIGURE 1. *Receptaculites canadensis* Billings' type. Fine-grained lens of dolomite in Cabot He' shale, Limehouse, Ont., Collector, R. Bell. G.S.C., Mus. No. 2500. (Page 35.)

FIGURE 2. *Synaptophyllum williamsi* Chadwick n. sp. Manitoulin dolomite east of Manitowani bay, Manitoulin island. G.S.C., Mus. No. 4508. (Page 128.)

FIGURE 3. *Acervularia gracilis* (Billings). Manitoulin dolomite, Manitowaning bay, Manitoul island. G.S.C., Mus. No. 4503. (Pages 30 and 31.)

FIGURE 4a. *Enterolasma* cf. *geometricum* (Foerste). Manitoulin dolomite, near Ice lake, Manitoul island. G.S.C., Mus. No. 4511. (Pages 30 and 31.)

FIGURE 4b. Calyx of 4a.

FIGURE 5. *Streptilasma* cf. *hoskinsoni* Foerste. Manitoulin dolomite, near Ice lake, Manitoul island. G.S.C., Mus. No. 4512. (Pages 30 and 31.)

FIGURE 6a, b; 7a, b. *Brockocystis tecumseth* (Billings). Manitoulin dolomite, near Ice lake, Ma toulin island. G.S.C., Mus. No. 4516, 4516a. (Pages 30 an 31.)

FIGURE 8. *Favosites cristatus* Edwards and Haime. St. Edmund dolomite, Centre bluff, Cal head. G.S.C., Mus. No. 4513. (Page 38.)

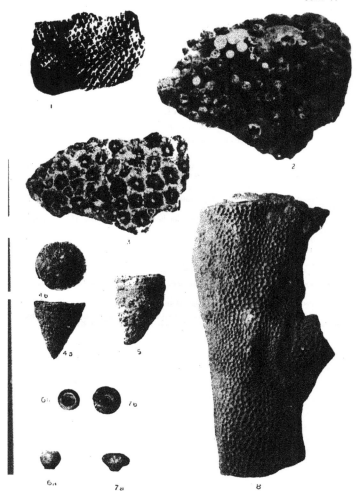

138

Plate VI.

140

PLATE VII.

Medinal ataract Fossils.

(Figures natural size.)

FIGURE 1. *Dalmanella eugeniensis* n. sp., pedicle valve. Manitoulin dolomite, Lavender falls, Ont. G.S.C., Mus. No. 4521. (Page 118.)

FIGURE 2a, b. *D. eugeniensis*, pedicle and frontal views of small specimen. Occurrence same as last. G.S.C., Mus. No. 4521b. (Page 118.)

FIGURE 3a, b, c, d. *D. eugeniensis*, pedicle, frontal, brachial, and cardinal views of small specimen. Cabot Head shale, Eugenia. G.S.C., Mus. No. 4522. (Page 118.)

FIGURE 4. *D. eugeniensis*, interior of pedicle and brachial valves, which are attached by rock. Manitoulin dolomite, Lavender falls. G.S.C., Mus. No. 4521b. (Page 118.)

FIGURE 5. *D. eugeniensis*, interior brachial valve. Cabot Head shale, 10 feet above Manitoulin dolomite, Eugenia. G.S.C., Mus. No. 4521. (Page 118.)

FIGURE 6. *D. eugeniensis*, interior pedicle valve. Locality same as Figure 5. G.S.C., Mus. No. 4522a. (Page 118.)

FIGURE 7a, b. *D. ampla eugeniensis* var. *subquadrata* n. var., pedicle and brachial valves. Cabot Head shale, Eugenia. G.S.C., Mus. No. 4604. (Page 118.)

FIGURE 8a, b. *D. eugeniensis* var. *pulvinata undata*, brachial and cardinal views. Cabot Head shale, Eugenia. G.S.C., Mus. No. 4604. (Page 118.)

FIGURE 9a. *Rhipidomella hybrida* (Sowerby), cardinal views, Cabot Head shale, Eugenia. G.S.C., Mus. No. 4617. (Page 35.)

FIGURE 9 b. Pedicle valve of 9a. (Page 35.)

FIGURE 10a, b. *Camarotoechia neglecta* (Hall), cardinal and frontal views. Cabot Head shale, Eugenia. G.S.C., Mus. No. 4521. (Page 35.)

FIGURE 11a, b, c. *Rhynchotreta cabotensis* n. sp., brachial, frontal, and pedicle views. Cabot Head shale at top of Dyer Bay dolomite, Clay cliff, near Cabot head. G.S.C., Mus. No. 4622. (Page 126.)

FIGURE 12. *Rhynchotreta lepida* Salvage, brachial view. Cabot Head shale just above Dyer Bay dolomite, Clay cliff, near Cabot head. G.S.C., Mus. No. 4626. (Page 35.)

FIGURE 13a, b. *R. cf. lepida*, pedicle and brachial views, small specimen. Locality same as Figure 12. G.S.C., Mus. No. 4626b. (Page 35.)

FIGURE 14. *R. cf. lepida*, a narrow specimen. Locality same as Figure 13. G.S.C., Mus. No. 4626a. (Page 35.)

FIGURE 15a, b. *Rhynchotreta gilliamae* Foerste, n. sp., pedicle and lateral views. Manitoulin dolomite, Manitowaning bay, Manitoulin island. G.S.C., Mus. No. 4331. (Page 126.)

FIGURE 16a, b. *Whitfieldella cataractensis* n. sp., brachial and lateral views. Cabot Head shale, near Owen Sound. G.S.C., Mus. No. 4627. (Page 121.)

FIGURE 17, a, b, c. *W. cataractensis*, brachial, pedicle, and lateral views. Cabot Head shale, Stoney Creek. G.S.C., Mus. No. 4628. (Page 121.)

FIGURE 18. *W. cataractensis*, pedicle view of another from Stoney Creek. G.S.C., Mus. No. 4628. (Page 121.)

FIGURE 19a, b, c. *Atrypa parksi* n. sp., brachial, pedicle, and frontal views. Cabot Head shale, Credit Forks. G.S.C., Mus. No. 4673. (Page 120.)

FIGURE 20. *Virgiana mayvillensis* Savage, pedicle valve. Dyer Bay dolomite, South bay, Manitoulin island. G.S.C., Mus. No. 4674. (Page 37.)

FIGURE 21. *Virgiana mayvillensis* Savage, pedicle valve. Dyer Bay dolomite, Tamarack cove, Manitoulin island. G.S.C., Mus. No. 4675. (Page 37.)

FIGURE 22a, b. *Lophospira pulchra* n. sp. Manitoulin dolomite, Cataract, Ont. G.S.C., Mus. No. 4678. (Page 123.)

FIGURE 23. *Strophostylus cf. cyclostoma* Hall. Manitoulin beds, Jordan, Ont. G.S.C., Mus. No. 4679. (Page 28.)

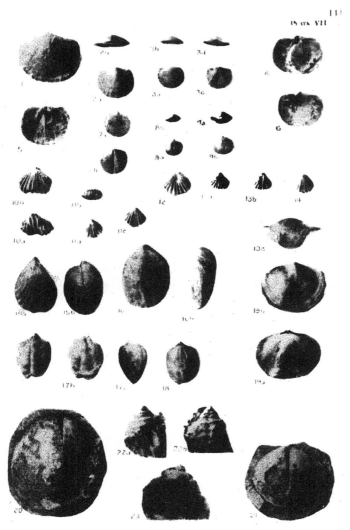

Plate VIII.

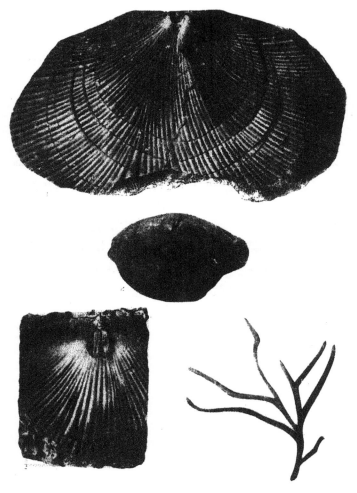

PLATE X.

Clinton Fossils.

(Figures natural size except 7b.)

FIGURE 1. *Spirifer radiatus* (Sowerby). Reynales (Wolcott) and Irondequoit dolomites. After Palæontology of New York, vol. II, pl. 22, fig. 2f. (Page 49.)

FIGURE 2. *Hyattidina congesta* (Conrad). 2a, brachial, 2b, frontal, views. Reynales (Wolcott) dolomite. After Palæontology of New York, vol. II, pl. 23, figs. 2c and 2d. (Page 49.)

FIGURE 3. *Cœlospira hemispherica* (Sowerby). 3a, pedicle, and 3b, brachial, valves. Sodus shale. After Palæontology of New York, vol. VIII, pt. II, pl. 82, figs. 2 and 3. (Page 48.)

FIGURE 4. *Whitfieldella cylindrica* (Hall). 4a, brachial, 4b, lateral, views. Irondequoit dolomite. After Palæontology of New York, vol. II, pl. 24, figs. 2a, 2c. (Page 51.)

FIGURE 5. *Whitfieldella intermedia* (Hall). 5a, brachial, 5b, lateral, views. Reynales (Wolcott) and Irondequoit dolomites. After Palæontology of New York, vol. II, pl. 24, figs. 3a, 3b. (Page 49.)

FIGURE 6. *Rhynchotreta robusta* (Hall), pedicle valve. Irondequoit dolomite. After Palæontology of New York, vol. II, pl. 23, fig. 7a. (Page 51.)

FIGURE 7. *Tentaculites minutus* Hall. 7a, specimens natural size; 7b, magnified about four times. Reported from "lower Clinton" of Clinton, N.Y., by Hall. After Palæontology of New York, vol. II, pl. 41a, figs. 8b, c. (Page 55.)

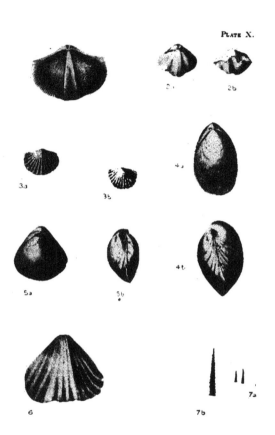

PLATE XI.

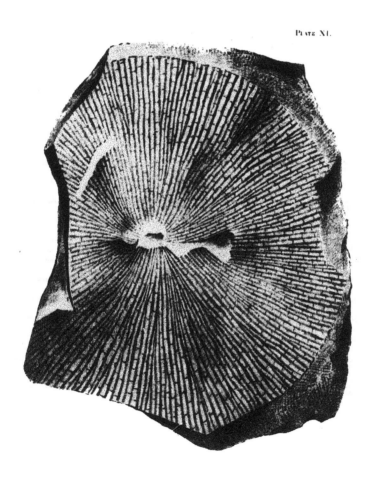

57237—12½

PLATE XII.

Rochester Fossils.

(Figures natural size except 10b.)

FIGURE 1. *Caryocrinites ornatus* Say. After Palæontology of New York, vol. II, pl. 49, fig. 1d. (Pages 54-56.)

FIGURE 2. *Stephanocrinus angulatus* Conrad. After Palæontology of New York, vol. II, pl. 48, fig. 1a. (Pages 54-56.)

FIGURE 3. *Spirifer cuspus* (Hisinger). 3a, 3b: two specimens from brachial side. After Palæontology of New York, vol. II, pl. 54, figs. 3c, 3d. (Pages 54-56.)

FIGURE 4. *Atrypa reticularis* (Linnæus). 4a, pedicle valve, 4b, frontal view. After Palæontology of New York, vol. II, pl. 55, figs. 5n, 5o. (Pages 54-56.)

FIGURE 5. *Atrypa nodostriata* Hall. 5a, brachial, 5b, lateral, views. After Palæontology of New York, vol. II, pl. 5b, figs. 2m, 2n. (Pages 54-56.)

FIGURE 6. *Camarotœchia obtusiplicata* (Hall). 6a, brachial, 6b, pedicle, views. After Palæontology of New York, vol. II, pl. 58, figs. 2a, 2b. (Pages 54-56.)

FIGURE 7. *Rhipidomella hybrida* (Sowerby). 7a, pedicle, 7b, lateral views. After Palæontology of New York, vol. II, pl. 52, figs. 4f, 4g. (Pages 54-56.)

FIGURE 8. *Dalmanella elegantula* (Dalman). 8a, pedicle, 8b, frontal, views. After Palæontology of New York, vol. II, pl. 52, figs. 3c, 3h. (Pages 54-56.)

FIGURE 9. *Whitfieldella nitida oblata* (Hall). 9a, lateral, 9b, brachial, view. After Palæontology of New York, vol. II, pl. 55, figs. 2e, 2a. (Pages 54-56.)

FIGURE 10. *Bilobites acutilobus* (Ringueberg). 10a, pedicle valve, 10b, the same enlarged. After Proc. Acad. Nat. Sc., Phila., 1858, pl. VII, figs. 5 and 5a. (Pages 54-55.)

FIGURE 11. *Whitfieldella nitida* (Hall). 11a, lateral, 11b, brachial, views. After Palæontology of New York, vol. II, pl. 55, figs. 1n and 1o. (Pages 54-56.)

FIGURE 12. *Spirifer niagarensis* (Conrad), pedicle valve. After Palæontology of New York, vol. II, pl. 54, fig. 5k. (Pages 54-56.)

FIGURE 13. *Spirifer niagarensis* (Conrad). 13a, brachial valve; 13b, interior pedicle valve, showing crura, (c). After Palæontology of New York, vol. VIII, pt. II, pl. 21, figs. 1 and 2. (Pages 54-56.)

Plate XII

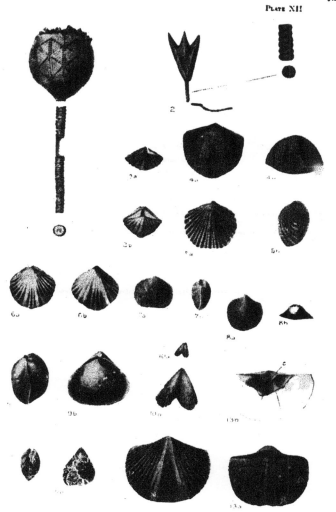

Plate XIII.

Rochester Fossils.

(Figures natural size.)

Figure 1. *Strophomena? decuvensis* n.sp., brachial valve. DeCew falls, Ont. G.S.C., Mus. No. 4687. (Page 124.)

Figure 2 a, b. *Schuchertella subplana* (Conrad), pedicle and cardinal views. From Palaeontology of New York, vol. II, pl. 53, figs. 10a, 10b. (Pages 54-56.)

Figure 3. *Platyceras angulatum* (Hall). After Palaeontology of New York, vol. II, pl. 60, fig. 1c. (Pages 54-56.)

Figure 4. *Homalonotus delphinocephalus* (Green). After Palaeontology of New York, vol. II, pl. 68, fig. 4. (Pages 54-56.)

Figure 5. *Calymene niagarensis* Hall. After Chicago Academy of Sciences, Bull. IV, pl. 23, fig. 10. (Pages 54-56.)

Figure 6. *Diaphorostoma niagarense* (Hall). After Palaeontology of New York, vol. II, pl. 60, fig. 1. Pages 54-56.

Figure 7. *Diaphorostoma hemisphericum* (Hall). After Palaeontology of New York, vol. II, pl. 60, fig. 2a. (Pages 54-56.)

Plate XIII.

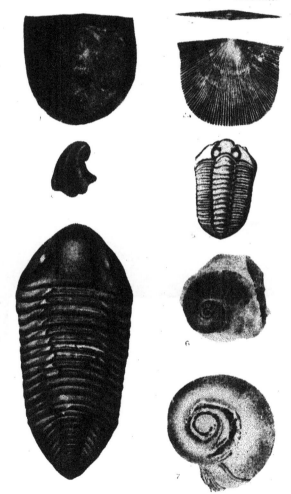

4

5

6

7

MICROCOPY RESOLUTION· TEST CHART

(ANSI and ISO TEST CHART No. 2)

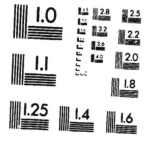

APPLIED IMAGE Inc

1653 East Main Street
Rochester, New York 14609   USA
(716) 482 - 300 - Phone
(716) 288 - 5989 - Fax

PLATE XIV.

Lockport Fossils.

(Figures natural size.)

FIGURE 1. *Zaphrentis stokesi* Edwards and Haime. Above Fossil Hill horizon, New England, Manitoulin island. G.S.C., Mus. No. 4688. (Page 65.)

FIGURE 2. *Zaphrentis stokesi* Edwards and Haime. Boulder bluff, Cabot head, lake Huron. G.S.C., Mus. No. 4694. (Page 65.)

FIGURE 3. *Alreolites labechei* Edwards and Haime. Skunk island, south side Manitoulin island. G.S.C., Mus. No. 2642a. (Page 65.)

FIGURE 4. *Ptychophyllum stokesi* Edwards and Haime. Fossil hill, Manitoulin island. G.S.C., Mus. No. 4696. (Page 65.)

PLATE XIV.

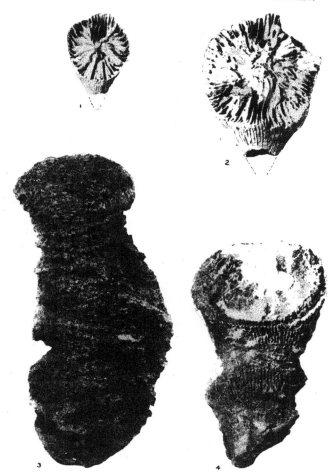

PLATE XV.

PLATE XVI.

Lockport Fossils.

(Figures natural size.)

FIGURE 1. *Synaptophyllum multicaule* Hall. Fossil hill. G.S.C., Mus. No. 5074. (Page 65.)

FIGURE 2. *Strombodes pentagonus* Goldfuss. Fossil hill. G.S.C., Mus. No. 5073a. (Page 65.)

FIGURE 3. *Syringopora dalmani* Billings' type. Lake Timiskaming, collected by W. E. Logan. G.S.C., Mus. No. 2618. (Page 65.)

PLATE XVII.

Lockport Fossils.

(Figures natural size.)

FIGURE 1. *Syringopora retiformis* Billings' type. Owen Sound, collector R. Bell. G.S.C., Mus. No. 2617. (Page 65.)

FIGURE 2. *Favosites hisingeri* Edwards and Haime. Fossil hill, Manitoulin island. G.S.C., Mus. No. 5075. (Page 65.)

FIGURE 3. *Paleofavosites asper* D'Orbigny. Fossil hill, Manitoulin island. G.S.C., Mus. No. 5138. (Page 65.)

FIGURE 4. *Syringopora fibrata* Rominger. Long bay, Manitoulin island, G.S.C., Mus. No. 5076. (Page 65.)

FIGURE 5. *Vermipora niagarensis* Rominger? Fossil hill, G.S.C., Mus. No. 5117. (Page 65.)

Plate XVIII.

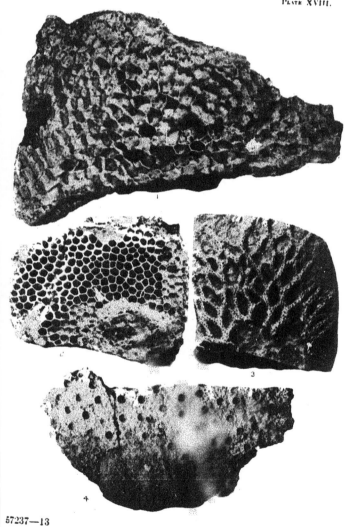

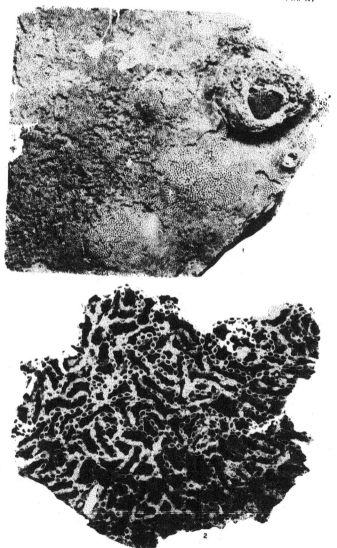

2

166

PLATE XX.

Lockport Fossils.

(Figures natural size.)

FIGURE 1. *Halysites catenularia microporus* (Whitfield). Fossil hill, Manitoulin island. G.S.C., Mus. No. 5125. (Page 65.)

FIGURE 2. *Stricklandinia manitouensis* n. sp. 2a, brachial, 2b, lateral, 2c, cardinal, views. Plateau east of Sandfield, Manitoulin island. G.S.C.,Mus. No. 5123. (Page 124.)

FIGURE 3. *Stricklandinia manitouensis* n. sp. Cardinal view of a comparatively thin specimen. G.S.C., Mus. No. 5126a. (Page 124.)

FIGURE 4. *Homœospira apriniformis* Hall. 4a, pedicle, 4b, frontal, views. Plateau east of Sandfield, Manitoulin island. G.S.C, Mus. No. 5127. (Page 67.)

FIGURE 5. *Spirifer eudora* (Hall). 5a, lateral view of a specimen from the Niagara group of Racine, Wisconsin; 5b, pedicle view of a specimen from the Niagara group of Waldron, Indiana. After Palæontology of New York, vol. VIII, pt. II, pl. 21, figs. 20, 21. (Page 67.)

FIGURE 6. *Pentamerus oblongus* Sowerby. Eight feet above base of Lockport dolomite at Wiarton. G.S.C., Mus. No. 5128. (Page 67.)

FIGURE 7. *Clorinda ventricosa* (Hall). 7a, brachial, 7b, lateral, views, of specimen from Niagara group of Waukesha, Wisconsin. After 20th Rept. New York Cabinet Natural History, pl. 13, figs. 18 and 19. (Page 67.)

PLATE XX.

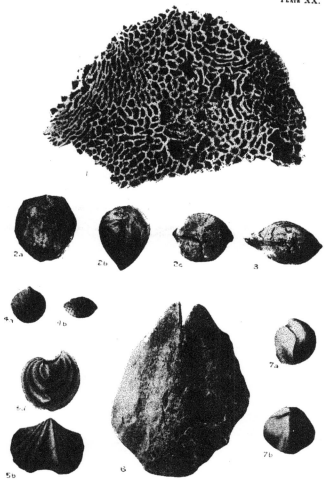

PLATE XXI.

170

PLATE XXII.

Guelph Fossils.

(Figures natural size.)

FIGURE 1. *Pycnostylus guelphensis* Whiteaves. Collected at Hespeler by E. Billings, 1857. G.S.C., Mus. No. 2789. (Page 77.)

FIGURE 2, *Pycnostylus elegans* Whiteaves type. Collected at Hespeler by T. C. Weston, 1867. G.S.C., Mus. No 2790c. (Page 77.)

FIGURE 3. *Pycnostylus elegans* Whiteaves type. Interior view, showing tabulæ and marginal septa. Collected at Hespeler by T. C. Weston, 1867. G.S.C., Mus. No. 2790. (Page 77.)

FIGURE 4. *Trimerella grandis* Billing. type. 4a, brachial, 4b, pedicle, views of an internal cast from Hespeler, Ont. G.S.C., Mus. No. 2803. (Page 77.)

FIGURE 5. *Whitfieldella hyale* (Billings). 5a, frontal, 5b, brachial, views. Hespeler, Ont. G.S.C., Mus. No. 2806. (Page 78.)

FIGURE 6. *Amphicœlia leidyi* Hall. From zinc prospect near Wiarton. G.S.C., Mus. No. 5129. (Page 78.)

Plate XXII.

PLATE XXIII.

Guelph Fossils.

(Figures natural size.)

FIGURE 1. *Megalomus canadensis* Hall. 1a, right valve, 1b, dorsal view. Collected at Elora by D. Boyle, 1880. G.S.C. Mus. No. 2827. (Page 78.)

FIGURE 2. *Megalomus canadensis* Hall. Right side of internal cast. From Galt, Ont. G.S.C. Mus. No. 5130. (Page 78.)

FIGURE 3. *Conchidium occidentalis* (Hall). 3a, cardinal area of cast of interior; 3b, brachial valve. After Geology of Canada, 1863, p. 337, fig. 341, a and c. (Page 77.)

FIGURE 4. *Mytilarca dentirostra* (Hall). Cast of interior. After 20th Rept. Cabinet of Natural History, New York State, pl. 14, fig. 2. (Page 78.)

PLATE XXIV.

Guelph Fossils.

(Figures natural size.)

FIGURE 1. *Liospira perlata* (Hall). 1a, top, and 1b, basal, views. From Galt, Ont. G.S.C. Mus. No. 2871. (Page 78.)

Plate XXIV.   175

1a

1b

PLATE XXV.

Guelph Fossils.

(Figures natural size.)

FIGURE 1. *Coelocaulus longispira* (Hall). Natural section showing interior. From Baptist harbour, Bruce peninsula. G.S.C., Mus. No. 5131. (Page 78.)

FIGURE 2. *Coelocaulus longispira* (Hall). Small specimen collected at Denham by J. Townsend, 1874-82. G.S.C., Mus. No. 2895a. (Page 78.)

FIGURE 3. *Hormotoma cf whiteavesi* Clarke and Ruedemann. This specimen is more slender than those described by Clarke and Ruedemann. Collected at Elora or Hespeler by T. C. Weston, 186.. G.S.C., Mus. No. 2859b. (Page 81.)

FIGURE 4. *Coelocaulus vittlia* (Billings). Collected at Galt by E. Billings, 1857. G.S.C., Mus. No. 2882. (Page 81.)

FIGURE 5. *Straparollina daphne* (Billings), umbilical view. Collected at Elora by D. Boyle, 18.. G.S.C., Mus. No. 2856. (Page 81.)

FIGURE 6. *Holopea guelphensis* Billings. Collected at Durham by A. Murray, 1857. G.S.C., Mus. No. 2848. (Page 79.)

FIGURE 7. *Euomphalus gallensis* Whiteaves. Collected at Durham by J. Townsend. G.S.C., Mus. No. 2853. (Page 78.)

FIGURE 8. *Murchisonia biltinijeana* Miller. Collected at Galt? by T. C. Weston, 1857. G.S.C., Mus. No. 2875. (Page 81.)

FIGURE 9. *Coelocaulus vitellia* (Billings). Collected at Durham by J. Townsend, 1882. G.S.C., Mus. No. 2881. (Page 81.)

FIGURE 10. *Trematonotus angustatus* (Hall). Collected at Hespeler by E. Billings and A. Murray, 1867. G.S.C., Mus. No. 2911a. (Page 80.)

Plate XXV

PLATE XXVI.

Guelph Fossils.

(Figures natural size.)

FIGURE 1a. *Orthoceras brucensis* n. sp. Partly natural and partly artificial longitudinal section. From Hay bay, Bruce peninsula, Ont. G.S.C., Mus. No. 5732. (Page 124.)

FIGURE 1b. Smallest camera showing position of siphuncle. (Page 124.)

FIGURE 2. *Poterioceras* n. sp., septum next to living chamber showing location of siphuncle. From lower Guelph at zinc prospect near Wiarton. G.S.C., Mus. No. 5133. (Page 79.)

FIGURE 3. *Poterioceras* n. sp., segment near distal end. G.S.C., Mus. No. 5133a. (Page 79.)

FIGURE 4. *Poterioceras* n. sp., much of living chamber and the greater part of the body. Locality same as Figures 2 and 3. G.S.C., Mus. No. 5134. (Page 79.)

180

PLATE XXVII.

Guelph and Cayugan Fossils.

(Figures natural size except 4 and 5.)

Guelph.

FIGURE 1. *Orthoceras brucensis* n. sp., showing external characters of shell. From Pine Tree harbour, Bruce peninsula, Ont. G.S.C., Mus. No. 5135. (Page 121.)

FIGURE 2. *Melonocrras arcticameratum* (Hall). From cape Hurd, Bruce peninsula, Ont. G.S.C., Mus. No. 5136. (Page 79.)

Cayugan.

FIGURE 3. *Eurypterus remipes* Dekay. From top of Bertie waterlime. After New York State Mus. Mem. 14, pt. 2, pl. 5, fig. 5. (Page 83.)

FIGURE 4. *Whitfieldella sulcata* (Vanuxem) magnified 2 diameters: pedicle valve. Akron dolomite one-quarter mile west of Ridgemount, Welland county. G.S.C., Mus. No. 5137. (Page 86.)

FIGURE 5. *Whitfieldella sulcata* (Vanuxem) magnified 2 diameters: frontal view showing sinus on both valves, the upper being the pedicle. G.S.C., Mus. No. 5137a. (Page 85.)

Plate XXVII.

PLATE XXVIII.

B. Medina sandstone and shale, from Manitoulin beds at bottom to Clinton dolomite at top of hill. Thorold sandstone is conspicuous bed near top. New York Central and Hudson River railway above Lewiston, New York. (Page 24.)

A. Lockport and Guelph dolomite. Thin beds 12 feet thick near centre are Eramosa beds. Below car barn, Gorge route, Niagara falls, New York. (Pages 62 and 73.)

PLATE XXIX

B. Section of Cabot Head shale, Grimsby sandstone and Reynales dolomite (at very top), at falls on Twentymile creek, south of Jordan. (Pages 32 and 39.)

A. Section of Clinton dolomite, Rochester shale, and lower part of Lockport dolomite, at Grimsby. (Pages 46 and 52.)

184

PLATE XXX.

A. Section of DeCew waterlime (a little more than the lower half of the picture) and the Gasport and overlying Lockport dolomite. Niagara gorge, New York. (Page 58.)

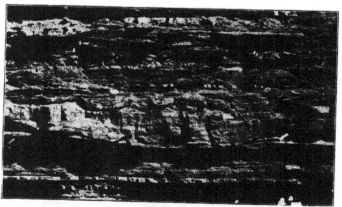

B. Section of upper Cabot Head shale showing current channel. Niagara gorge, New York. (Page 34.).

A.   Bay dolomite and underlying red Cabot Head shale, at a locality 3 miles west of Cabot head. (Page 33.)

B. Quarry in Manitoulin dolomite, Owen Sound. (Page 28.)

186

Plate XXXII.

B. Niagara escarpment. Cliffs of Lockport dolomite near Kolapore, Grey county. (Page 57.)

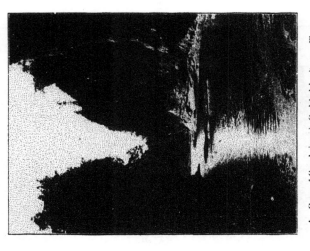

A. Gorge of Grand river, in Guelph dolomite near Elora (Page 72.)

Plate XXXIII.

A. Eramosa beds and base Guelph dolomite, near Guelph. (Page 62.)

B. Upper Lockport and lower Guelph dolomite (caves below contact), Half-way rock, north end Bruce peninsula. (Page 62.)

PLATE XXXIV.

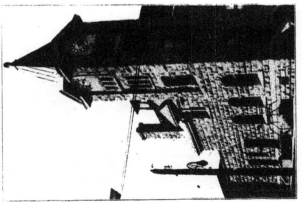

B. Larger Flower pot, composed of Eramosa dolomite, capped with Guelph dolomite. Flowerpot island. Georgian bay. (Page 62.)

A. Shelburne post-office, made from local Lockport dolomite. (Page 97.)

INDEX.

A.

B.

C.

V.

W.

Y.

Z.

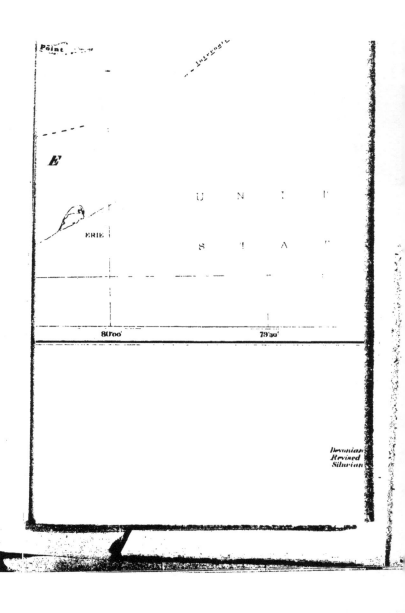

Point

E

ERIE

U N I
S A

80°00'   79°30'

Devonian
Revised
Silurian

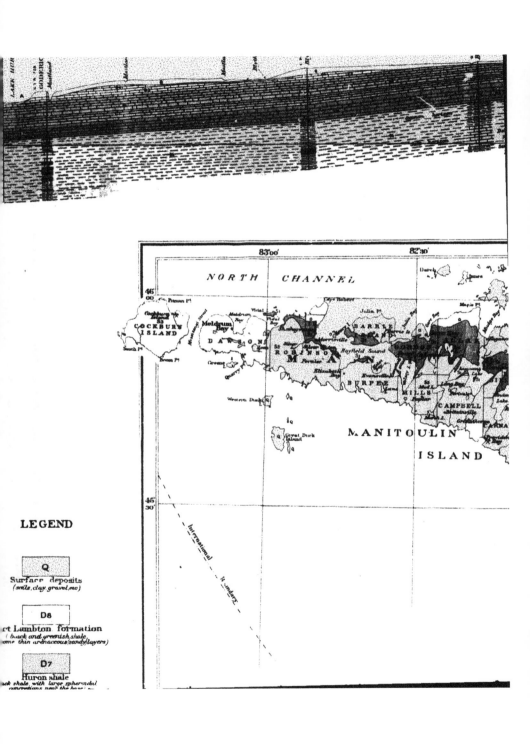

LEGEND

Q

Surface deposits
(soils, clay, gravel, etc.)

D8

rt Lambton formation
(back and greenish shale,
ome thin arenaceous (sandy) layers)

D7

Huron shale
ack shale, with large spheroidal
concretions near the base,

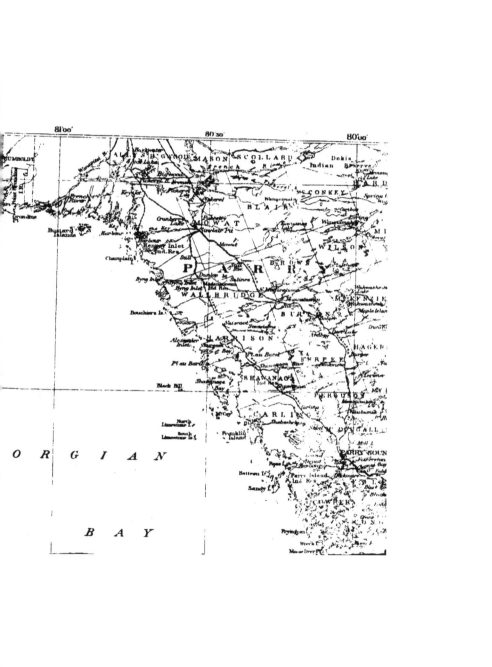

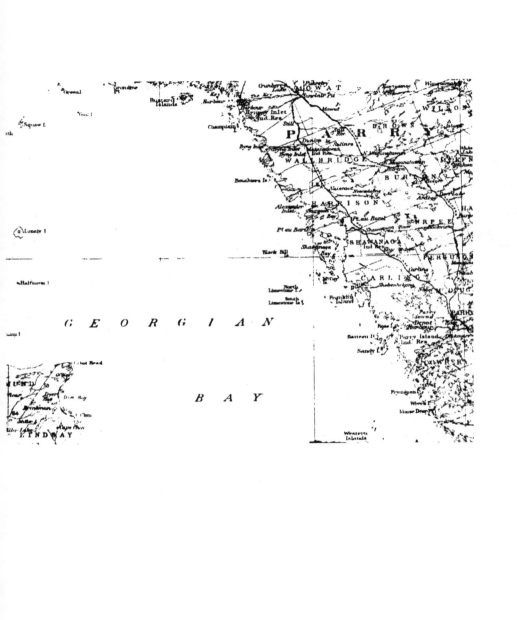

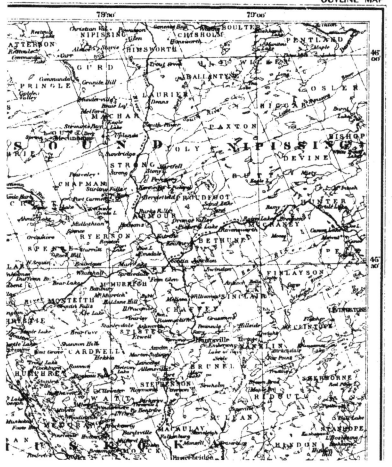

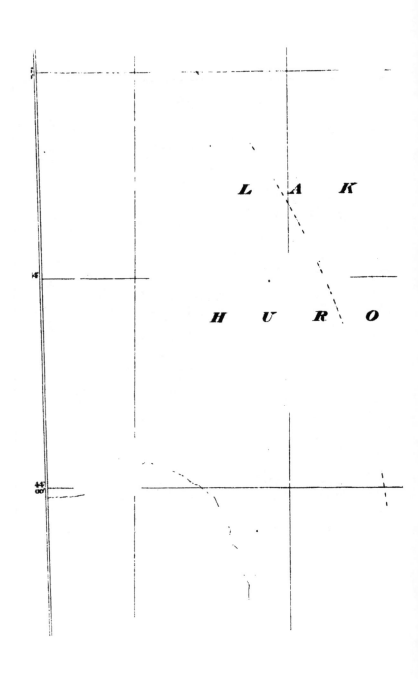

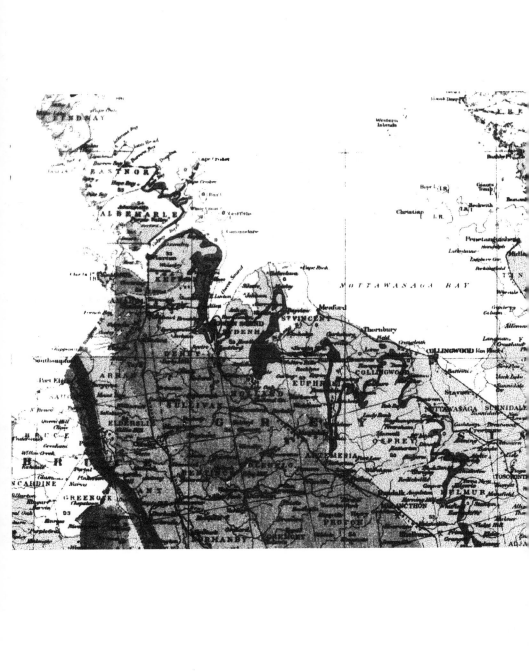

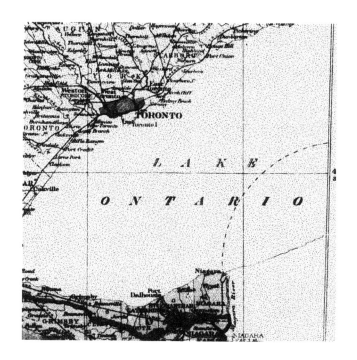

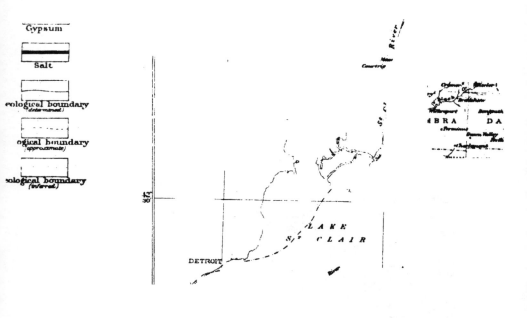

Gypsum

Salt

ological boundary
(determined.)

ogical boundary
(approximate.)

ological boundary
(inferred.)

St. C.

River

River
Courtrig

dBRA DA

Orchard   Market

Black   Bradshaw

Walkerport   Bungeuth

Dawn Valley
North

atXacigaignard

42
30'

DETROIT

LAKE
St. CLAIR

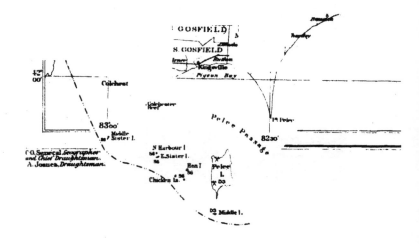

GOSFIELD

S. GOSFIELD

Ruthen

Kingsville

Pigeon Bay

Ranglah

Ranglay

42
00'

Colchest

Colchester
River

Pelee Passage

1st Pelee

83 00'

Middle
Sister I.

N Harbour I.

E. Sister I.

Hen I.

Pelee
L.

82 30'

C.O. Senecal *Geographer*
and Chief Draughtsman.
A. Jomes, *Draughtsman.*

Chicken Is.

Middle I.

near by M.Y. Williams

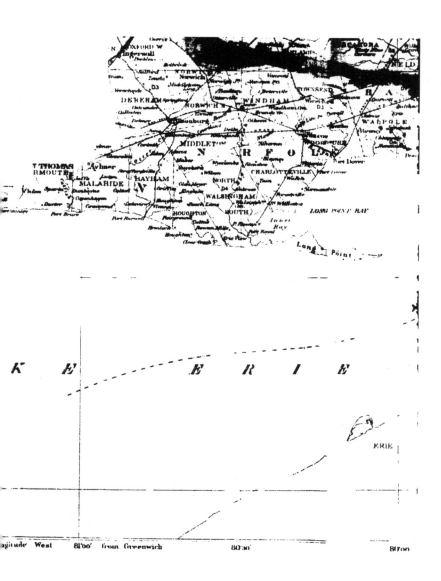

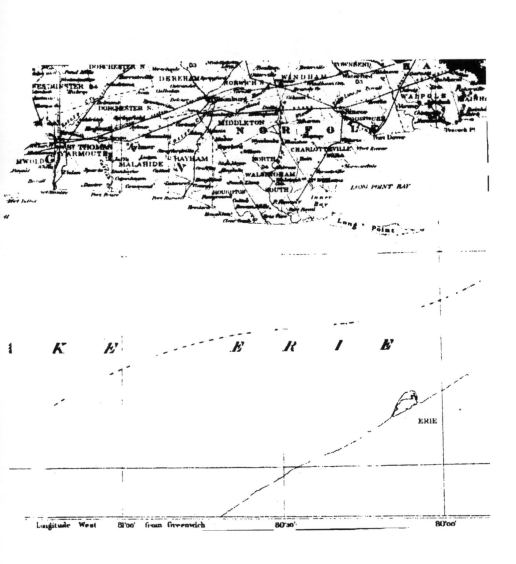

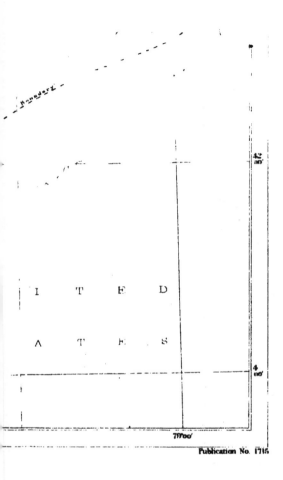

Boundary

4.2
30'

I   T   E   D

A   T   E   S

4
00'

77°00'

Publication No. 1715

Devonian geology by C.R. Stauffer, 1910-1912.
Revised by M.Y. Williams, 1918.
Silurian geolo   M.Y. Williams, 1912-1916.

Publication No. 1715

*Base map from engraved plates
of the Department of the Interior*

*Devonian geology by C.R.Stauffer 1910 -1912
Silurian geology by M.Y.Williams 1912 -1916*

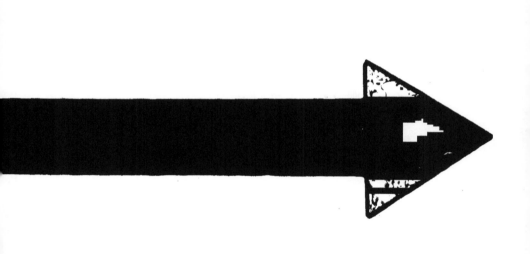

MICROCOPY RESOLUTION TEST CHART

(ANSI and ISO TEST CHART No. 2)

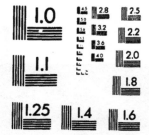

APPLIED IMAGE Inc

1653 East Main Street
Rochester, New York   14609   USA
(716) 482 - 0300 - Phone
(716) 288 - 5989 - Fax

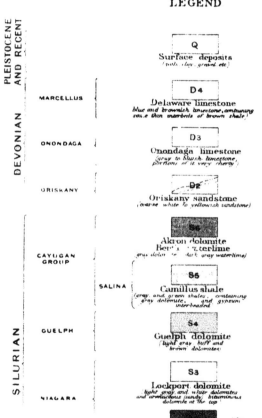

LEGEND

PLEISTOCENE AND RECENT

Q
Surface deposits
(soils, clay, gravel, etc.)

DEVONIAN

MARCELLUS

D4
Delaware limestone
(blue and brownish limestone, containing some thin interbeds of brown shale)

ONONDAGA

D3
Onondaga limestone
(gray to bluish limestone, portions of it very cherty)

ORISKANY

D2
Oriskany sandstone
(coarse white to yellowish sandstone)

SILURIAN

CAYUGAN GROUP

S6
Akron dolomite
Bertie waterlime
(gray dolomite to dark gray waterlime)

SALINA

S5
Camillus shale
(gray and green shales, containing gray dolomite, and gypsum interbedded)

GUELPH

S4
Guelph dolomite
(light gray, buff and brown dolomites)

NIAGARA

S3
Lockport dolomite
(light gray, and white dolomites and arenaceous (sandy) bituminous dolomite at the top)

Lightning Source UK Ltd.
Milton Keynes UK
UKHW02f0827100918
328635UK00012B/671/P